Electrochemical Processes and Corrosion in Reinforced Concrete

Some reinforced concrete structures prematurely corrode as they age, with significant financial implications, but it is not immediately clear why some are more durable than others. This book looks at the various mechanisms for corrosion and how what seemed to be a relatively simple matter has become more complex the further it is understood due to the properties of concrete, steel and the way reinforced concrete structures are constructed. The significance of electrochemical processes is identified with recent research using new technology discussed. It is third volume in this CRC Focus program by this author with the other titles 'Cathodic Protection for Reinforced Concrete Structures' and 'Durability of Reinforced Concrete Structures'.

Specialist contractors, consultants and owners of corrosion damaged structures will find this an extremely useful resource. It will also be a valuable reference for students at postgraduate level.

Paul Chess is currently the Managing Director of Corrosion Mitigation Limited and a part time lecturer at Sussex University. He was formerly the Managing Director of the largest specialist manufacturer of products for cathodic protection of concrete in the world.

Electrochemical Processes and Corrosion in Reinforced Concrete

Paul Chess

CRC Press is an imprint of the
Taylor & Francis Group, an informa business

CRC Press
Boca Raton and London
First edition published 2024
by CRC Press
4 Park Square, Milton Park, Abingdon, Oxon, OX14 4RN

and by CRC Press
6000 Broken Sound Parkway NW, Suite 300, Boca Raton, FL 33487-2742

British Library Cataloguing-in-Publication Data
A catalogue record for this book is available from the British Library

Library of Congress Cataloging-in-Publication Data
Names: Chess, Paul, author.
Title: Electrochemical processes and corrosion in reinforced concrete / Paul Chess.
Description: First edition. | Boca Raton : CRC Press, 2023. | Includes
bibliographical references and index.
Identifiers: LCCN 2023001574 | ISBN 9781032392417 (hbk) |
ISBN 9781032392424 (pbk) | ISBN 9781003348979 (ebk)
Subjects: LCSH: Reinforcing bars—Corrosion. | Reinforced
concrete—Corrosion. | Steel—Corrosion.
Classification: LCC TA445.5 .C484 2023 | DDC 620.1/37—dc23/eng/20230131
LC record available at https://lccn.loc.gov/2023001574

ISBN: 978-1-032-39241-7 (hbk)
ISBN: 978-1-032-39242-4 (pbk)
ISBN: 978-1-003-34897-9 (ebk)

DOI: 10.1201/9781003348979

Typeset in Times
by codeMantra

Contents

1 Electrochemical processes **1**
1.1 Introduction 1
References 10

2 Concrete and steel composition **11**
2.1 Description of concrete 11
2.2 Flow in porous materials 13
2.3 Steel rebar 14
2.4 The steel-concrete interface (SCI) 17
References 19

3 The relationship between chloride level and corrosion **21**
3.1 Introduction 21
3.2 Laboratory research on corrosion threshold level and thereafter corrosion rate 24
3.3 Surveys on real structures for a chloride corrosion threshold level and active corrosion rate 25
3.4 Review of corrosion threshold level 26
3.5 The effect of the structures shape on corrosion 27
3.6 The effect of chloride on pitting corrosion 28
3.7 Conclusions 28
References 29

4 Ionic solvation and transport **31**
4.1 Ionic movement through water without a potential difference 31
4.2 Ionic movement through water with a potential difference 32
4.3 Ionic movement through concrete with a potential difference 35
4.4 Changes to the steel with the flow of ions caused by a potential difference 36
4.5 Chloride movement through reinforced concrete with cracks 38
References 39

5 Passivation and depassivation **41**
5.1 Introduction 41
5.2 Film formation 41
5.3 Breakdown of the film 45
References 46

6 Electrochemical theory **47**
6.1 Introduction 47
6.2 Reversible potential and the single electrode 47
6.3 Electrochemical cells 48
6.4 Reversible electrodes and the Nernst equation 48
6.5 The potential pH diagram 51
6.6 Tafel Plot and Evans Diagram 51
6.7 Combining reduction and oxidation in an Evans diagram 53
6.8 Polarisation curves 54
6.9 Linear polarisation resistance 56
6.10 Acoustic emission 56
6.11 Conclusions 56
Reference 57

7 Electrochemistry reality of steel in concrete **59**
7.1 Introduction 59
7.2 Tafel slopes measured of steel in simulated pore solutions (SPS) 60
7.3 Tafel slopes measured of steel in concrete 60
7.4 Corrosion rates recorded in active and passive steel in reinforced concrete 61
7.5 Passive and active corrosion potentials measured in reinforced concrete structures 62
7.6 Modelling for life expectancy and reality 64
7.7 Non-destructive testing for corrosion of steel in concrete 65
7.8 Conclusions 66
References 67

8 State of the art **69**
References 72

Index 73

Electrochemical processes

1

1.1 INTRODUCTION

Metals have been of critical importance in the development of the human world order and the table below shows the correlation in society evolution in concert with the development of new materials. The adoption of bronze and iron were so important that they are given the titles 'age'. The existence of metal ores has been of great significance in the development of countries. For example, the industrial revolution in England where large scale production of metals started had several causations but the existence of a relatively pure iron ore and plentiful supply of metallurgical coal was of critical importance along with a newly available class of engineering entrepreneurs. As shown in Figure 1.1 there was a slow development of engineering materials which has accelerated dramatically in recent times which reflects the increased pace of industrialisation throughout the twentieth century and beyond.

The use of electrochemistry in industry became widespread about 150 years ago with batteries becoming widely used from medical research to early speed record-breaking cars. Fuel cells were also invented in this time period. It was also being used for industrial processes such as chrome and copper plating. The manufacture of elements by electrical or electrochemical techniques such as aluminium and chlorine were all in widespread use before the turn of the twentieth century. All these processes or objects were developed and optimised with very little, or no theoretical input from electrochemical engineering principles. The level of innovation and optimisation that was achieved by simple trial and error is remarkable. For example, in the process of chromium plating it was found by raising the temperature they could change the density of the deposits allowing a hard chrome layer to be achieved. They also engineered the anode and cathodes for different effects and changed the electrolytes viscosity to obtain a better finish.

DOI: 10.1201/9781003348979-1

Date	material	location
9000 BC	wrought copper	middle east
3000 BC	copper smelting	middle east
2500 BC	silver and alloys	middle east
2000 BC	bronze age	far east
1500 BC	iron age-wrought	middle east
600 BC	cast iron	china
1300 AD	cast iron	europe
1740	cast steel	england
1838	copper electroplate	england
1884	aluminium	france
1910	titanium	USA
1913	stainless steel	england
1933	hastelloys	england
1963	carbon fibre	england

FIGURE 1.1 Evolution of materials development.

Over the last hundred years electrochemical processes, have been, and are being used on an ever widening list of processes which are important in producing metals, coatings, non-metallic compounds and offer great hope of a greener future. Electrochemical processes are central to future hydrogen production and many types of fuel cells (Figure 1.2) for producing electrical power already exist. This coupled with the massive increase in the use of batteries for almost every purpose are great examples of the positive benefits of electrochemical reactions. What is not perhaps fully appreciated is the role of electrochemistry in biological processes. Electrochemical energy is produced in every plant and animal. For example, an animal nervous system send its signals by means of electrochemical reactions. Virtually every electrochemical process and its technological application has a role in modern medicine.

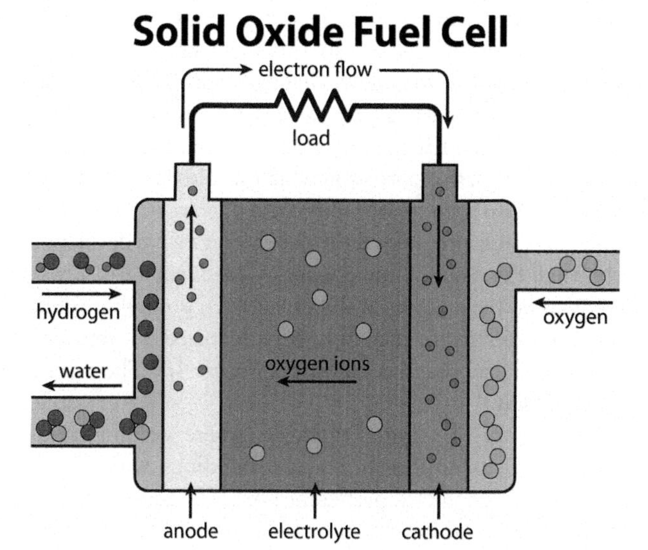

FIGURE 1.2 Anatomy of a fuel cell showing its electrochemical operation.

In general, electrochemical principles have been developed in laboratories on simplified and optimised experiments which bear little comparison to the extremely complex electrochemical processes occurring in nature and to a lesser extent in industry.

Despite the huge economic importance of electrochemistry, it presently does not comprise a large part of a chemistry syllabus for university undergraduates. For example, take a book titled Atkins (2014), Physical Chemistry which is sometimes considered the definitive British text book for university undergraduate chemistry students. Electrochemical processes are only addressed on 27 pages of a 959 pages book. Further, physical chemistry is only one component of the three main topics which are normally studied in a chemistry degree which also includes inorganic and organic components. The processes that are described in Atkins are a quick run over of electrochemical processes which are idealisations where little or no mention is made of the non-idealised nature of actual electrochemical processes or the poorly understood factors in its behaviour.

Despite all the benefits to humanity of metals and intended electrochemical processes, unchecked electrochemistry has a downside in that it is the main mechanism for the corrosion of metals. The other is high-temperature corrosion which can, for example, occur on the blades of a gas turbine. The corrosion process can be in many forms but generally the material degrades back to its oxide or similar form losing its mechanical and other desirable properties. This problem with engineering materials has been understood for many thousands of years particularly since the advent of the iron age. There is evidence that, in concert with early smelting, iron was being protected by animal fats to prevent active corrosion which would degrade the functioning of the implement. Also seen in archaeological digs was coating with oils, tallow or bitumen. As time progressed further coatings were developed and have been widely used. For example in a paper by Blackney (2017), wrought iron railings mounted on the external walls of a church ground in Bath, England, were installed in 1765 and to date have been recoated 37 times. The first 35 times by traditional lead paints as shown in Figure 1.3. The railings are still in excellent condition.

In the same way of animal evolution where certain creatures have remained successful for millions of years, buildings which have survived for long periods have been of a particular type. The most durable structures constructed by man so far have been constructed of masonry. For example, in Figure 1.4 the Colosseum was built of travertine limestone, tuff and brick-faced concrete with wood and fabric also being incorporated in 80 AD. The wood and cloth are long gone but despite earthquakes and stone thieves the majority of the structure is still standing. Incidentally, the pozzolanic nature of the cement used in this concrete and similar structures seems to impart a durability not replicated in modern structures.

Composite building structures where the varying properties of different materials have been combined have been used for many thousands of years. This has mainly been stone in compression with wood used for tensile stress elements. If the wood was kept dry then its life span could be dramatically increased and most buildings reflect this in their design such as rafters being overlaid with rooftiles. It was noted quite early on that where movement of masonry was occurring then cramps could be used to prevent collapse of the structure. Some standard cramp designs commonly used in Greek and Roman era structures are shown in Figure 1.5.

For some reason man has desired to make larger and larger structures. For many centuries the most important and largest structures had a religious purpose. In the west these buildings were commonly cathedrals. As these structures grew larger, taller and more elaborate they required structural additions to relieve tensile forces. This coupled with the increasing quantities of metals being smelted allowed their use on a large scale in structures. An example of this is St Paul's cathedral in London which is shown in Figure 1.6. Here a 100 m wrought iron chain was wrapped around the base of the dome to counter outthrust of the dome. This chain was considered insufficient or had badly corroded and a 'non-rusting' metal band

FIGURE 1.3 Various paint coatings on railings to prevent corrosion of the iron.

FIGURE 1.4 Roman colosseum built in 80 AD.

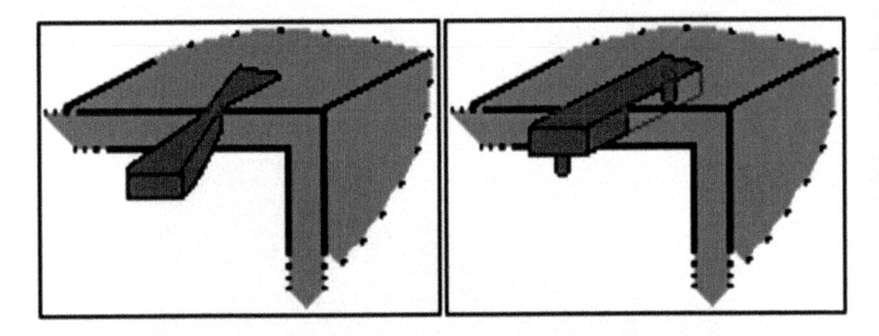

FIGURE 1.5 Bronze cramp designs commonly used for Roman and Greek structures.

FIGURE 1.6 St Paul's cathedral with wrought iron chain wrapped around the base of the dome completed in 1711.

was applied in 1930 to make the original chain redundant. The specification of a 'non-rusting' material suggests that there was significant corrosion of the original chain. The use of a stainless steel would increase the material price by a factor of six or more over carbon steel with the same mechanical properties at this time.

Further advancements in smelting processes allowed new and improved metals and alloys to be developed and become widespread. The nineteenth century saw a great increase in the use of metals, particularly varieties of iron-based products. Later the use of cast and particularly wrought iron decreased while steel in the twentieth century has become the predominant engineering material.

A new form of cementituous material called ordinary portland cement (OPC) was invented early in the nineteenth century and was selected by Joseph Bazalgette for the London Super Sewer project which was completed by 1875 using 670,000 cubic meters of concrete and 320 million bricks affixed with this material. OPC has gone on to become globally the most popular cement due to its many advantages and low financial cost. Cements and aggregates (stones of various sizes) are mixed together to produce concrete.

Due to both steel and concrete's particular strengths and weaknesses it was inevitable that they would be combined to produce a composite material. These characteristics are that concrete is quite strong in compression, cheap, mouldable, good fire resistance and durable. Steel has it all as a metal in that is cheap, strong in both compression and tension, isotropic, good fire resistance, similar expansion rate to concrete and excellent fatigue properties. Its only significant drawbacks are its limited corrosion resistance and high density. The corrosion resistance of steel can be dramatically improved by metallic additions to the alloy, but this also increases the cost factorially.

In 1853 the first use of iron reinforcement in a concrete matrix was undertaken and this usage advanced so that by the turn of the twentieth century, steel reinforcement was being used in large civil structures. By the 1920s some researchers led by Atwood (1924) had noted that some reinforced concrete structures were suffering from corrosion of the reinforcement, particularly in marine areas. The use of reinforced concrete exploded after the second war with massive new capital projects using this composite material. At this time the general consensus in the civil engineering industry was that this material was effectively everlasting with zero maintenance. Design life was being steadily increased until 120 years was specified on many occasions and this is still the case in many national and international standards. Perhaps this design life is achievable in some structures, but it was also becoming increasingly apparent that, in certain situations, significant corrosion damage to the steel reinforcement was occurring over a short time period which was only a small percentage of the design life. Initially this damage was treated by knocking out the cracked and delaminated concrete, but this did not stop the problem which kept recurring over larger areas as the exposure time increased. Some research on the causes of this problem was undertaken but this research effort was much increased when post Second World War built reinforced concrete bridges were found to be suffering problems. This was because there were national research centres for designing and improving highway structures in existence so sustained research programmes could be funded. The biggest individual research programme was undertaken by the Americans, who had become increasingly concerned about the poor durability of their road infrastructure and the massive costs in repair. The first SHRP (Strategic Highways Research Program) was started in 1985 and had a budget of $150 million over five years to look at the problems and come up with solutions. After this time a second program was initiated called SHRP2 with substantial further funding to undertake further research on durability solutions. One of the most important topics which was studied in these programs was the cause of premature reinforcement corrosion in highways structures and what to do about in repair. Later some thought was given to specifying new structures that would not have these problems, but there was limited progress on this as it involved an increased initial capital cost of structures which was strongly resisted by all sides of the industry and the USA is still plagued by poor corrosion performance of its reinforced concrete infrastructure.

In this book the material concentrated on is carbon steel with concrete as this is by far the most commonly produced and used metal composite. Other metals have been used for specific applications and non-metallics are now also being used as reinforcement in a few concrete structures. Steel is used in massive quantities to reinforce concrete structures in almost every country of the world. According to the WSO (World Steel Organisation) (2020) the total world steel production for 2019 was 1,900 million tonnes with 51% of this used for buildings and infrastructure. Of this amount about 44% (this may seem a low percentage but also includes all steel bridges, electricity pylons, wind turbines etc.) of this was used as rebar so by a rough calculation 426 million tonnes of steel reinforcement was used in one year. Typically the mass of steel reinforcements is about 2% of a reinforced concrete structures weight, so in the region of 21.3 billion tonnes of construction in a single year can be estimated as the world output of steel reinforced concrete. The sheer quantity of the material used illustrates the economic importance of ensuring that these structures are as durable as possible. There is an interesting subset to the use of steel and this is in the form of post tensioned structures. Here ducts of either galvanised steel tubing, or more latterly plastic tubes, are placed in the structure with steel tendons inside. After casting and setting these tendons are typically stressed up to 70% of their yield point. In older installations the steel tendon was often un-grouted and coated with grease but now it is more commonly grouted. The reason these are mentioned is that these are commonly used in high performance and high-cost structures such as long span bridges or nuclear containment vessels as shown in Figure 1.7. These are highly stressed structures with a criticality not to fail and yet are

FIGURE 1.7 Nuclear power station suffering significant corrosion to its two post tensioned containment domes which are the structures close to the water.

not likely to follow the same corrosion mechanisms discussed in this book. On the other hand, pretensioned steel structures may be expected to behave in a similar way as a conventional reinforced concrete structure from a corrosion point of view.

Corrosion science as distinct to corrosion empirical testing was probably started by Faraday in 1834 who discerned a quantitative connection between corrosion weight loss and electrical current generation. Further corrosion principles were established at a surprisingly early time, for example, in 1900 the Nernst equation was published with the Tafel equation in 1905 and then the Debye-Huckel theory of electrolytes. Whitman and Russell (1924) observed that both the anodic and cathodic reactions were simultaneously occurring when steel was actively corroding. The mixed potential (which is the combined anodic and cathodic potential) was a 1930's refinement with the Pourbaix potential -pH diagrams coming in the 1940s. These early studies were looking at establishing rules for both the thermodynamic principles (which is if a reaction can happen) and to a certain extent kinetic principles (the rate that a reaction does proceed). In almost all the early laboratory experiments attempts to idealise the experiments were made. So selected pure metals and very dilute solutions were the normal. The progress of the experiments was monitored by techniques such as titration. Galvanostats and potentiostats were not available until the 1950s so more specific research was limited with varying small currents or voltages. These 1920's laboratory principles were further defined by Ulick Evans (1937) who has had probably the greatest individual influence on academic corrosion advancement as he grasped the critical role of passivation and its converse in depassivation. He made many important advances and one of his many volumes 'is still considered the most comprehensive book written by one man on corrosion and protection'. The present state of the corrosion science art is as mentioned in 'Corrosion' (Shrier 1994) which with multiple authors is probably the most authoritative current text book on all aspects of this topic. They state that 'it should be appreciated that the use of a particular metal or alloy in a given environment is based usually on previous experience and empirical testing rather than on the application of scientific knowledge – the technology of corrosion is without doubt in advance of corrosion science and many of the phenomena of corrosion are not fully understood.'

To explain the cause of the originally unexpected corrosion of steel reinforcement in concrete, elements of corrosion science were used to analyse the data obtained from laboratory experiments. Early on simplified conditions and analogue solutions were used which was very difficult to relate to data obtained from real structures. These laboratory tests have generally proven to be unable to give transferable results. This has not been helped by the limited time frames of experiments which require that artificial corrosion inducing processes are introduced. As has become more apparent with time there are complex and various conditions pertaining in an actual reinforced concrete structures with different pieces of steel, different depths of cover, various diffusion processes, advection processes, multiple saturation levels and other influences all having various effects on the degree and rate of corrosion of the steel reinforcement. This has made it extremely difficult to model or predict the structural life with any degree of precision as will be discussed later.

REFERENCES

Atkins P, de Paula J, *Atkins physical chemistry*, 9th edition, Oxford University Press, ISBN 9780199697403, 2014

Atwood W, Johnson A, The disintegration of cement in seawater, Transactions, 1924, ASCE, 87, 204–230

Blackney K, www.buildingconservation.com, painting historic ironwork, 2017

Evans U, *Metallic corrosion passivity and protection*, E. Arnold and Company, 1937

Shrier L, Jarman R, Burstein, *1:14 Corrosion,* Butterworth-Heinemann, ISBN 075061077 8, 1994

Whitman W, Russell R, Effect of hydrogen ion concentration on the submerged corrosion of steel, *Industrial Engineering Chemistry Part 1*, 1924, 16, 7, 665–670

WSO, World Steel Organisation, worldsteelorganisation.org, 2020

Concrete and steel composition

2

2.1 DESCRIPTION OF CONCRETE

Concrete is a composite material which is full of pores which has an intrinsic effect on its behaviour. Concrete comprises a mixture of cement and aggregates. The cement is a material with adhesive and cohesive properties which make it capable of bonding the mineral fragments into a compact whole. Most of the cement types used for construction can set and harden under water and are thus called hydraulic cements. There are three main categories of cements commonly used in construction; namely natural (in the trade called pozzolans), Portland and high alumina (this has fallen out of favour in the last few decades). The most commonly used around the world is Portland and this is discussed more fully here though from a corrosion perspective all three should behave in a broadly similar manner. A typical Portland cement comprises a majority of calcium oxide with silicon oxide and other oxides, iron oxide typically is about 3% of the mass of the dry cement. This cement is then hydrated by adding water during the concrete casting process, this produces solid products of hydration, gel water and empty capillary pores. As the cement ages typically these capillary pores will be filled with gel water and more solid products will be created. Mature cement paste has capillary pores which form an interconnected system randomly distributed throughout the matrix. These capillaries can get filled with gel. The gel itself has porosity of typically 28% and these pores are of various sizes. In hardened cement several forms of aqueous solutions have been identified, namely free water, gel water and bound water. These are the constituents of hydrated compounds which are typically saturated solutions.

DOI: 10.1201/9781003348979-2

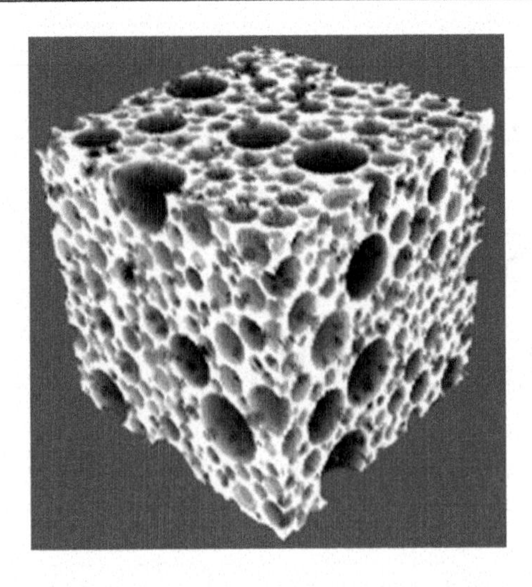

FIGURE 2.1 Idealised and magnified concrete microstructure showing size range and distribution of pores.

Typically, three quarters of the volume of concrete is occupied by aggregates which is normally used in two sizes namely fine and coarse aggregate. The aggregate is usually from crushed stone of many types. Each of the aggregate types will also have certain porosity. For example, limestone has a porosity of between 0% and 37.6% (Figure 2.1) as described by Neville (1995). The boundary between the cement and aggregate is of significant importance in the behaviour of concrete. This area is called the Interface Transitional Zone (ITZ) and is presently believed to be the most important highway for the flow of porous materials such as chloride ions. Shane (2000) has stated that its effective diffusivity is between 6 and 12 times the level of bulk cement.

A critical parameter of the concrete is its porosity connection which is a measure of how much the individual pores are connected through both the cement paste and the aggregate.

It should be noted that in civil engineering increased compressive strength is often conflated with the likely durability of a mix. In reality the permeability and ductility of the concrete are often of much greater importance to vulnerable structures service life then its compressive strength. Ductility is rarely measured in brittle materials such as concrete as it is very hard to do in a laboratory and even worse on site. It is known however that concrete additives such as silica fume while reducing the diffusivity of the bulk concrete also reduces the ductility and thus, in certain situations, increases the chances of cracking. This factor probably accounts for its poorer performance on actual structures than would be anticipated by its laboratory performance as reported by Gjorv (2014) in Section 3.3 and Luping (2012) in Chapter 7.

2.2 FLOW IN POROUS MATERIALS

Concrete in common with most inorganic construction materials (such as brick or stone) has open porosity. It thus takes up water according to its exposure to rain, groundwater, condensation and humid air and releases it through the drying effects of the atmosphere. Until recently the main focus had been the flow of liquid water or ions in saturated materials and the transport of water vapour in relatively dry materials (Figure 2.2).

Recently it has become recognised by Hall (2012) that unsaturated flow is the most important water transport mechanism in a porous material. In some reinforced concrete structures it is possible that there could be different transport mechanisms on a daily cycle. For example, a bridge support column in a tidal marine location could be immersed at high tide causing saturation and air exposed at low tide allowing partial saturation after solar drying. Unsaturated flow is controlled by hydraulic processes which are strongly influenced by chemical forces on the pore surfaces. This influences the binding of the water against the pore wall. The water travels as a capillary flow through the unsaturated concrete along these pore surfaces. When low-quality concrete with a high water content is used in structures it is common to see preferential corrosion on the underside of the rebars. This is because there can be a substantial direct flow at the rebar to concrete junction in this location. This is also seen with plastic spacer strips which are used for impressed current cathodic protection anodes on new structures and post tensioning plastic ducting (it should be noted that the galvanised ducting behaves much better in this regard as it slowly reacts to form a bulky zinc oxide film which blocks this ITZ). In these circumstances the flows can be very high

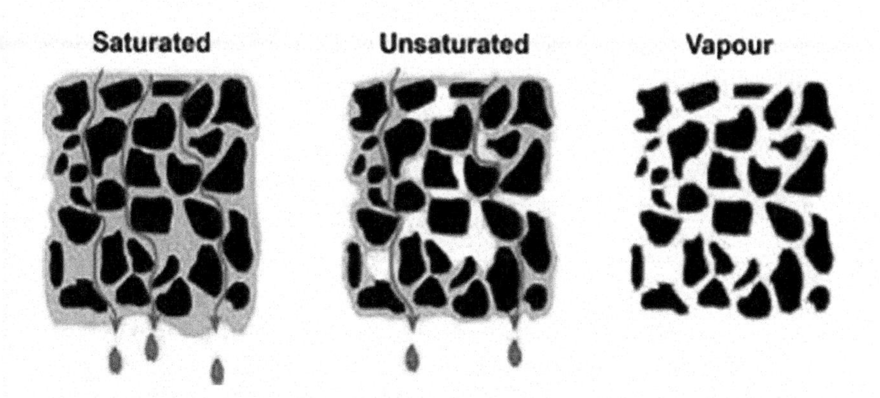

FIGURE 2.2 Examples of the three water flow conditions in concrete.

relative to calculations based on concentration diffusion. If this flow is transporting water from a fresh source on one side to a dry area on the other side of the structure then it is not likely to have deleterious consequences, but where there is salt water (sabkha for example, which is an Arabic term for saline ground water) at below ground level and a hot and dry environment above ground level this unsaturated flow can cause severe and premature corrosion. This is one of the main reasons for the unexpected and poor durability of reinforced concrete structures in the Arabian Gulf. Advection is the process where dissolved ions are carried by this capillary flow processes. This means dissolved ions can be transported long distances much more quickly then would be predicted by saturated flow calculations.

The pore sizes found in concrete are significantly smaller than the pore sizes found in soils. Water is drawn from material with large pores to material with small pores. This means that the concrete is commonly saturated even though the surrounding earth seems fairly dry. In foundations this can mean that the concrete is saturated and then capillary action forces water upwards. Because there is little or no evaporation below ground level the structure is typically saturated right up to ground level. As concrete structures do not have a damp-proof course this water continues to rise. In locations where there is significant evaporation the final height this water reaches is defined by the wetting branch of the water retention curve. What is critical here to durability is what is deposited or concentration enhanced from the ground level upwards by this process. If it is chloride it will have a fundamental importance for long-term durability as in this above ground location there is a high oxygen availability and possibly high temperatures. This capillary action process is the same as that used by trees. Here water is drawn out of the soil by this difference in pore size, taken up xylem pipes and transpired in reactions in the leaves. This transpiration process causes a negative pressure. Dissolved salts in the soil water are used to make parts of the tree. The tree structure is optimised for this capillary process but to get an idea of scale an American Redwood will pump 2,200 L of water to a height of 100 m each day.

2.3 STEEL REBAR

Steel is a composite material made of iron and carbon with other minor elements. The small size of the carbon atom allows it to enter the iron as interstitial solute atoms rather than as a substitutional solid solution which is what happens with larger atoms such as chromium. The carbon atoms put strain

FIGURE 2.3 Rebar after installation and before concrete casting.

on the iron lattice and this dramatically improves the mechanical properties compared to ferrite which is essentially pure iron (Figure 2.3).

The steel used for rebar can be cold rolled which improves the mechanical properties and is still used for some high-performance reinforced concrete structures but is today fairly rare. Almost all rebar available now is hot rolled where the steels final shape is achieved when at a high temperature. As a result of this hot rolling process the actual surface of as manufactured rebar is mill scale which is an oxidised form of the iron thus a certain form of rust. The next level in of the rebar is a tempered martensitic layer. This a hard layer which is relatively brittle with the carbon not forming a separate compound and being retained in the iron crystal due to the rapid cooling of the outside of the rebar in the production process. Further in from the rebar surface there is slower cooling. This allows interlocking plates of ferrite and cementite to form. This is commonly called pearlite as it looks like mother of pearl under a microscope. Cementite is very hard and brittle and essentially behaves as a ceramic. Ferrite is very soft. This pearlite can be thought of a composite material which provides the desirable properties of both toughness and strength (Figure 2.4).

From a durability point of view the cementite and ferrite have different electrical potentials so you will get very small electrochemical cells between them. If steel is cleaned to bright metal and placed in a saline solution these

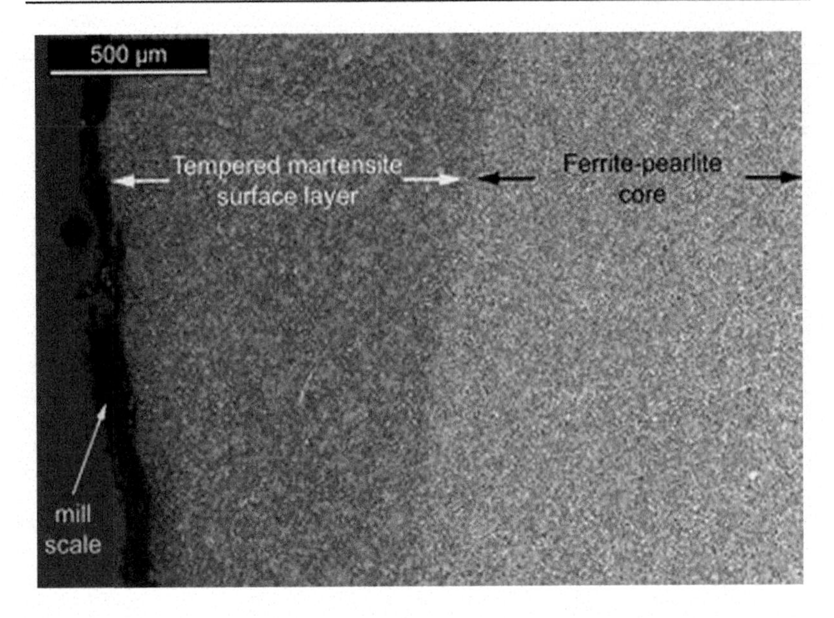

FIGURE 2.4 Hot worked rebar showing the actual surface interface with the concrete.

small electrochemical cells cause a phenomena known as flash rusting. This is an orange/brown film being formed minutes after exposure.

Flash rusting commonly occurs on grit blasted steel as shown in Figure 2.5. This corrosion process will not be possible to measure in a traditional corrosion experiment as the anode and cathode areas are too small to measure individually.

Early rebar up until 1930 or so was smooth and commonly square in cross section. To improve the bond strength this square section rebar was then twisted. The present ribbed pattern was generally adopted post the Second World War. The earlier cold rolling process has being superseded by hot rolling in the majority of cases. Post the Second World War to the turn of the century most rebar was sourced from virgin materials using a blast furnace. Since the turn of the century in Europe rebar has been primarily produced with scrap steel using electric arc furnaces and this is also common in the USA. This scrap steel has a different chemical composition to primary steel due to elements such as copper, zinc and others being present. This is because, for example, copper-cored wiring is laced through items such as cars which provide the metal being reprocessed. It may also have a different microstructure as the steel is continually cast (concast) rather than bloomed. This in combination with a change away from OPC in large projects to GGBS

FIGURE 2.5 Flash rusting after grit blasting steel.

(granulated blast furnace slag) and other cements might cause a modern structure to display a different durability to an identical structure in the same environment constructed 30 years ago. There is some evidence observed by Gjorv (2014) and discussed in more detail in Chapter 7 that this has occurred and supposedly higher-performance modern materials are showing the same or lower durability than their predecessors.

2.4 THE STEEL-CONCRETE INTERFACE (SCI)

The steel to concrete interface is important to the bond strength between the two materials but perhaps of greater importance is its affect on the corrosion behaviour. Originally the steel to concrete interface is bounded by mill scale. This scale appears to be relatively stable in concrete and has been detected on the surface of the rebar after several decades of exposure in a corrosive environment. Normally the mill scale is separated from the steel surface by other corrosion products. The original mill scale film has many cracks and crevices so it does not provide an ideal barrier coating so additional corrosion products accumulate under and around it as shown in Figure 2.6a and b.

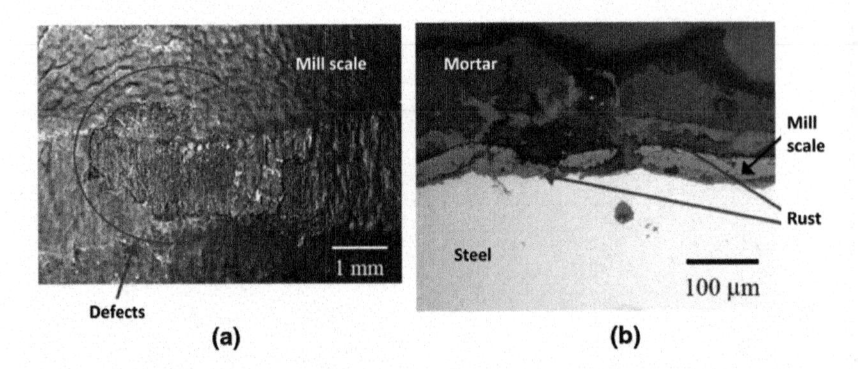

FIGURE 2.6 Micrographs showing different oxidation products in addition to mill scale (a) additional oxidation product beside the millscale (b) additional oxidation product both over and under the millscale.

There is always a lead time from manufacture of the rebar to concrete pour and this may have a significant effect on the surface corrosion product on the rebar. This could include chloride containing compounds if the rebar is transported or stored in a marine atmosphere.

When the rebar is surrounded by concrete in the casting process, the alkalinity of the cement causes a gradual change to the oxide surrounding the actual metal surface at cracks and other areas where direct contact can be made around the mill scale. This appears to happen much more slowly than has been found with calcium hydroxide analogue solutions. An analogue solution is intended to be chemically similar to concrete pore water.

There is presently significant ignorance about the corrosion behaviour of the steel in the SCI and it is likely that many of the laboratory experiments with analogue solutions and cleaned rebar may, at best, be indicative of what is occurring in real structures. It is apparent from examination of groups of older structures of a similar design in similar environments, such as bridges, tunnels and jetties that they show similar corrosion degradation behaviour despite being built in different countries with diverse design codes and construction practices. Using this information it should be possible to devise more realistic laboratory analogues in the future. A further problem for realistic laboratory experiments is the corrosion initiation time allowance on laboratory trials compared to that in a real structure. This sounds relatively innocuous but actually represents a significant problem as even in the most extreme cases structures are exhibiting the first signs of damage after ten years exposure. This is much longer than most research programmes can be funded for. This means that accelerating the laboratory programme is required by techniques such as pre-rusting, chloride dosing or electrical charging. Any of these could cause a significant disparity between the trial and reality.

REFERENCES

Gjorv O, *Durability designs of concrete structures in severe environments*, CRC Press, ISBN 97814665 87298, 2014, p30–31

Hall C, Hoff W, *Water transport in brick stone and concrete*, ISBN 978 0415 564670, 2012

Neville A, *Properties of concrete*, Longman, ISBN 0582 230705, 1995, p129

Shane J et al, Effect of the interfacial transition zone on the conductivity of Portland cement mortars, *Journal of the American Ceramic Society*, 2000, 83(5), 1137–1144

3 The relationship between chloride level and corrosion

3.1 INTRODUCTION

From the early days after reinforced concrete was first used as a construction material it was noted that in marine atmospheres corrosion of the rebar was occurring. This was reported by Atwood (1924). In this particular study more than 3,000 references were assembled. At most inland locations this degradation process has not been observed but on some specific structures such as bridges and multi-storey car parks in northern latitudes they have been prematurely damaged. Here the problem was caused by chlorides which are used for the purpose of de-icing. In the Arabian Gulf the chloride is present in the ground water and advection moves it to the reinforcement as described in Section 2.2 (Figure 3.1).

Other degradation mechanisms such as alkali aggregate reaction (AAR), freeze-thaw, chemical attack and carbonation have all been found on reinforced concrete structures but these problems are dwarfed by the problem of corrosion of the steel reinforcement and will not be considered further.

It has been a long held contention that there has to be a certain chloride level in the concrete where corrosion of the steel will start at a measurable or meaningful rate. This appears on the surface to be a reasonable idea and is

DOI: 10.1201/9781003348979-3

FIGURE 3.1 Apartment block in Dubai scheduled for demolition ten years after construction because of reinforcement steel corrosion.

likely to be correct for a specific size and type of steel in a specific alkaline solution but this does not seem to be correct for real structures. Many studies over the last 50 years have been made ranging from laboratory studies with analogue electrolytes to sampling actual structures where corrosion is deemed to be occurring at a significant rate with the underlying aim of establishing a specific level of chloride where active corrosion begins. Indeed, this belief has been so pervasive that the concrete compositions in both national and international standards have a maximum level of chloride which is permitted. A typical example limit is 0.4% by weight of cement on a reinforced concrete construction with lower chloride levels of 0.1% allowed in prestressed or post-tensioned structures as described in BS 8500:2019. The logic for the lower limit in the prestressed case is presumably the higher vulnerability of these structures but the reasoning is not explained.

In principle, the critical chloride level at which the corrosion becomes appreciable can be coupled with the time to initiation, then the rate of corrosion to estimate the service life of the structure. So here you have date of initiation and then a corrosion rate. Unfortunately for this simple and apparently reasonable premise, there are massive uncertainties in this approach which we will discuss in more detail later but can be outlined as; when will corrosion start occurring, will the corrosion rate stay constant or accelerate as cracking occurs and finally what is the damage that can be tolerated by the structure. For these reasons this approach which has been commonly used to design life structures needs to be abandoned or at least used in concert with other factors. The damage that can be tolerated has typically been thought of as when the structures mechanical behaviour or bending capacity is reduced to a critical level, but this is generally not the case in the actual design life. For example with two extremes; a fuelling jetty in Gosport, England, was structurally assessed to be able to cope with 100% loss of rebar section in some locations and significant concrete spalling into the water with no loss in functionality and so it was allowed to continue to operate despite monthly spalling of concrete and up to 70% loss of section of the rebar. Spalling was the end of life reason for the Roosevelt Bridge illustrated in Figure 3.2 as it was feared concrete could fall onto the foot and cycle path below the main structure. At the Guggenheim art gallery in New York a light iron oxide staining of small parts of the façade concrete caused by minimal corrosion with no discernible degradation of the rebar was enough to replace this part of the structure and apply cathodic protection to other parts of the structure which were deemed to have a minor risk of future corrosion but were visible to the public.

Over the last 50 years, whole series of experiments in almost every industrialised country into the relationship between chloride level and corrosion of rebar in concrete or an analogue have been made. These and particularly real

FIGURE 3.2 Roosevelt Bridge, Florida, closed after 24 years due to corrosion of the reinforcement.

structure investigations have also been made which has gradually disclosed that the whole picture is much more complex than previously imagined. This research has shown that the concrete composition and structure, steel composition and surface layer, exposure conditions and structural geometry all play significant or even dominant factors in whether active corrosion is occurring. The testing can be split into two categories.

3.2 LABORATORY RESEARCH ON CORROSION THRESHOLD LEVEL AND THEREAFTER CORROSION RATE

These laboratory tests can be further split into three broad categories. Those with an aqueous electrolyte, sand and finally mortar/concrete. The most commonly used is an aqueous alkaline solution (typically sodium or calcium hydroxide) usually dosed with either sodium or calcium chloride. This type of test is reproducible and can give relatively quick data but does not replicate the actual likely transport mechanisms (see Chapter 4) and also probably does not replicate the chemical interim reactions observed on real structures during the corrosion process. The use of semi-saturated sand dosed with a chloride solution is likely to be closer to the transport mechanism occurring in concrete with the great advantage that you can remove the steel specimen (as you can with the aqueous electrolyte) and visually inspect it at various intervals in the test. This removes one of the most intractable problems, which is how do you measure accurately micro-corrosion from a remote location to the reinforcement. Steel in mortar or concrete is widely used and is the most realistic analogue though normally the concrete is significantly dosed with chloride so that a reasonable experimental time frame is achievable. An even dosing of chlorides is unlikely to occur in most structures unless added as set accelerator which has not been common for the last 30 years. Normally the aggregate size is significantly reduced to reduce the variability of the specimens. These experiments have allowed the effects of alkalinity, chloride/hydroxide ratios, cement types, aggregate types, steel rebar composition and mill scale to all be noted on the corrosion process and there is some value in these experiments but there are still several problems, including measuring micro-corrosion such as pitting without destructive examination of the specimens which then finishes the experiment.

3.3 SURVEYS ON REAL STRUCTURES FOR A CHLORIDE CORROSION THRESHOLD LEVEL AND ACTIVE CORROSION RATE

Systemic surveys have typically been on bridges. For example, Vassie (1984) looked at bridges in England which had been damaged by de-icing salts. Stratfull (1975) in the USA looked at bridges with marine environment induced damage in the South, notably in Florida which due to its Keys has the most bridges in the country, and de-icing salt damage on bridges in the North of America. The methodology of both the surveys was similar looking for physical damage such as cracking and iron oxide staining from the reinforcement, plus they also took the electrochemical potential of the steel from the concrete surface in these damaged areas. This data was then correlated against concrete dust samples at the cover depth which were titrated to obtain a chloride level. Their findings were remarkably similar in 0.25–1.5% by weight of cement for Vassie and 0.2–1.4% for Stratfull for the critical chloride content to initiate discernible corrosion. Vassie was fully aware of the Americans work in this area as there had been a British tour of the USA previously to understand their efforts in documenting and fighting reinforcement corrosion in highway structures and he adopted a similar survey technique. The bridges surveyed were typically constructed in the early 1960s and had about 20 years in service with the design life being 120 years for the UK and 75 years for the USA (Figure 3.3).

Gjorv (2014) has looked closely at reinforced concrete jetties around Norway's coast. The temperature variations and exposure conditions of all these jetties was reasonably similar. It was found that the most important factor in whether corrosion was occurring was not the exposure time or the chloride level but the design of the structure. To illustrate this Gjorv detailed a 60 years old structure having no corrosion despite high chloride levels. This was contrasted with an 8-year-old cruise liner terminal in Trondheim made with 380 kg of OPC and 5% silica fume per cubic meter of concrete. There had been a 0.1% chloride penetration to 35 mm with an average cover of 50 mm. Most of the deck support beams on the entire structure were actively corroding though there were no visible signs of corrosion on the surface.

In another investigation by Gjorv (2014) in Oslo harbour on a reinforced concrete jetty the lower deck beams had deteriorated and he noted 'these observations demonstrate how effectively the corroding steel in the lower part of the deck beams had functioned as sacrificial anodes, and thus cathodically

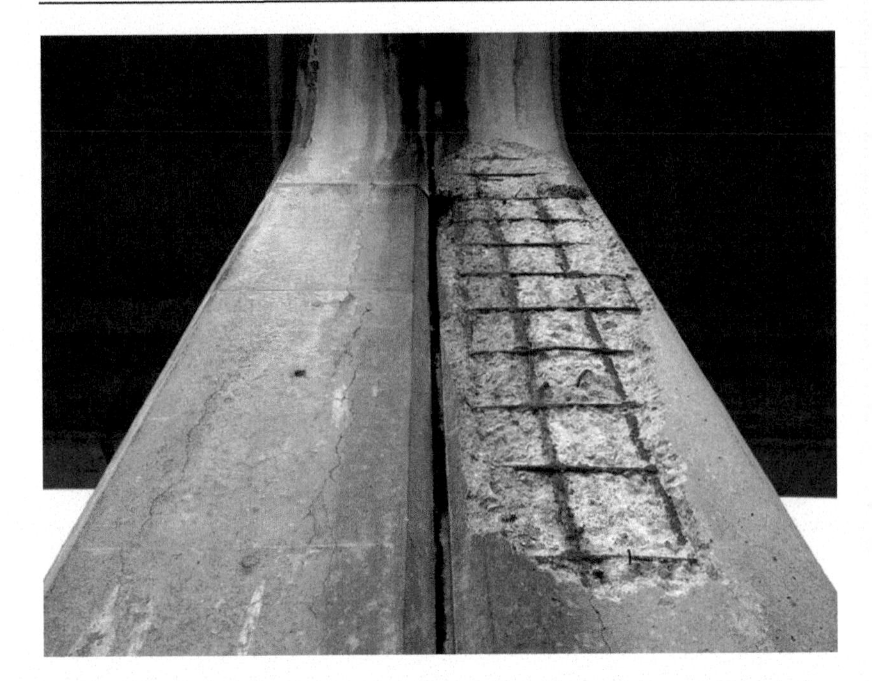

FIGURE 3.3 De-icing salt leaking through a deck expansion joint and damaging the columns. The column on the left side shows concrete surface damage and the column on the right the cause of this damage. This structure has been in service for 35 years.

protected the rest of the rebar system in the deck'. An early study by Arup (1995) also mentioned the importance of the electrochemical potential on corrosion initiation of real structures.

3.4 REVIEW OF CORROSION THRESHOLD LEVEL

In a review of this subject by Angst (2009) huge differences in chloride levels required to initiate corrosion were found in real structures. The range was from 0.1% to 1.95% chloride by weight of cement which is a 1,950% difference. What this means is that the presence of chloride at any practically measurable level may initiate corrosion but the presence of large amounts of chloride does not necessarily mean that corrosion will occur and it is thus other factors which determine whether this initiation and propagation phase happen. It was noted by Angst (2009) in this review of the critical corrosion initiation threshold that electric fields had a more dominant effect on whether the reinforcement corrodes then all the other variables, such as cement type or chloride concentration.

3.5 THE EFFECT OF THE STRUCTURES SHAPE ON CORROSION

Recently an experimental program was undertaken by Angst (2017) where various reinforced concrete specimens of different sizes were trialled. All other variables were kept the same for the tests. It was found that samples with reinforcement of respectively 1 cm, 10 cm and 100 cm length required very different chloride concentrations to initiate corrosion. No corrosion was found at more than 2.4% chloride by weight of cement for the 1 cm sample. The 10 cm sample started corroding at an average of 1.5% chloride by weight of cement. The 100 cm sample started corroding at an average of 0.9% chloride by weight of cement.

This experiment has profound implications in that it demonstrates that the steel reinforcement layout has a dramatic effect on the results obtained both in experimental procedures and also more importantly in real structures. Also, there is the implication is that large real structures are likely to perform substantially worse than may be predicted by laboratory testing with small samples. The reason for this behaviour was conjectured to be that the bigger the area of the steel reinforcement the more likely there was to be inhomogeneities on the surface of the steel where this anodic reaction can prosper.

Whatever the reasons for the results this experiment showed that the presence and layout of steel in the concrete has a dramatic effect on the corrosion process of steel in concrete and this should be considered highly important. Most of the laboratory durability experiments that have been undertaken to date have neglected this factor though researchers in the field such as Gjorv (2014) in Section 3.3 had noted geometric factors had a huge impact on the corrosion process. The three-dimensional construction of all reinforced concrete cages ensures that there is likely to significantly different potentials in different areas of the reinforcement which is likely to promote the generation of an anode and a cathode cell promoting corrosion. In general corrosion, the relative sizes of the anode and cathode is very important in the severity of the corrosion process. For example, having a large cathode and small anode induces a very high corrosion rate. There is some evidence that this is also true in reinforced concrete structures as seen by Chess (2009).

3.6 THE EFFECT OF CHLORIDE ON PITTING CORROSION

Chloride-induced reinforcement corrosion can be a highly localised process occurring at discrete and small locations on the rebars surface. This has been traditionally very difficult to study with traditional analytical techniques. A technique called laser ablation inductively coupled plasma mass spectroscopy or LA-ICP-MS has been recently developed which allows chemical analysis of very small areas (0.1–1 mm^2). This technique has been used by Silva (2013) to study in greater detail the role of chloride ions at the concrete to steel interface. It has been found that when active corrosion was occurring along the surface of the interface that there is a considerable variation in chloride concentration with higher concentrations around active sites. This was in contrast to steel in a passive state where the concentration of chloride along the concrete to steel interface was relatively homogenous. From these findings it was suggested that the chloride ions were actively being moved by an electrochemical potential gradient between the anode and cathode. These active anodes and cathodes have been observed to be caused by differences in the structure of the steel (such as mill scale or manganese sulphide inclusions) or on the concrete side of the interface, air voids. Once pitting corrosion was successfully activated Silva reported that the reaction kept occurring with its rate possibly controlled by the formation of hydroxyl ions at the cathode (cathodic control) or movement of oxygen and chloride through the pit (anodic control). This research ties in well with the practical findings of Gorv and Arup in that a potential gradient is driving ions on the steel to concrete interface and initiating active corrosion. It should be noted however that the finding that once the pit is initiated it stays active is contradicted by Angst (2017) who found that in most cases corrosion was initiated and then ceased for reasons not presently understood.

3.7 CONCLUSIONS

There is no doubt that the presence of chloride containing ions at the level of the steel to concrete allows a dramatic increase in the oxidation rate of the steel. This has been apparent for a long time in almost all reinforced concrete structures in at risk locations, such as marine or where de-icing salts are present. The notion that a certain chloride concentration to initiate corrosion appears reasonable and could still be partially true but has been very difficult to ascertain using dust samples on real structures and previous generations of analytical procedures. A new generation of test equipment such as LA-ICP-MS will allow studies in detail of much smaller areas in the future in laboratories. Unfortunately, this advance will be very difficult to replicate when studying real structures. What has been found recently is that if a limited amount of chloride is present then the chloride level is not the most dominant factor in initiating corrosion. The most important factor appears to be the electrical field difference between an anode and a cathode. This electric field may be sufficient to move chloride ions to the anode on the steel surface concentrating them to a sufficient extent to initiate pitting.

REFERENCES

Angst U et al, Critical chloride content in reinforced concrete-a review, *Cement and Concrete Research*, 2009, 39, 1122–1138

Angst U et al, The steel -concrete interface, *Materials and Structures*, 2017, 50, 143

Arup H, Sorensen H, A proposed technique for determining chloride thresholds, Proceedings of chloride penetration into concrete, International RILEM workshop, Saint Remy-les-Chevreuse, France, 1995, p460–469

Atwood W, Johnson A, The disintegration of cement in seawater, *Transactions ASCE*, 1924, 87, 204–230

BS8500:2019 Concrete – Complementary standard to BS EN206

Chess P, Drewett J, Cathodic protection of reinforced concrete swimming pools, *Concrete solutions*, CRC Press ISBN 9780429206603, 2009, p11–17

Gjorv O, *Durability designs of concrete structures in severe environments,* CRC Press, ISBN 97814665 87298, 2014, p30–31

Silva N, Chloride induced corrosion of reinforcement steel in concrete, PhD thesis, Chalmers University of technology, Gothenberg, Sweden, 2013

Stratfull R, Jurkovich W, Spellman D, Corrosion testing of bridge decks, *Transportation Research Record*, 1975, 539, 50–59

Vassie P, Reinforcement corrosion and the durability of concrete bridges, *Proceedings of the Institute of Civil Engineers Part 1*, 1984, 76, 713–723

Ionic solvation and transport 4

4.1 IONIC MOVEMENT THROUGH WATER WITHOUT A POTENTIAL DIFFERENCE

Early research in the nineteenth century looked at the moment of gas through diffusion, where the material moved from a higher to a lower pressure volume through a porous wall or a small tube. If the process consists of molecular flow rather than bulk flow through an orifice this is called effusion. Measurement of the effusion rates found that at a constant temperature and pressure drop the rates of effusion were inversely proportional to the densities and hence sizes of the gas molecules and this has become known as Graham's Law of Effusion.

Using this information as a basis the early idea was that electrolytes (acids, bases or salts) are more or less disassociated in an aqueous solution, or at least at very low concentrations known as infinite dilution. In this case the ions that are present act as free independent particles in a manner which is analogous to the motion of a gas and it was reasoned that the size of the molecules or ions would again determine their diffusion rate (Figure 4.1).

This seemed a reasonable contention but as further data was collected it was realised that this was too simple a model for solvated ions. Even at infinite dilution there needed to be some consideration of the interaction of the ions with the solvent and interaction of the ions with each other. Of course, as the concentration of the ions increased the interactions would be likely to increase until the saturation state where they are very significant factors for the movement of ions in an aqueous environment.

The fact that the movement of these ions does not follow Graham's law has been explained by various works around such as Stokes radius, degree of solvation and some newer ones such as solvation sheath. These are all used to try and explain why the ionic mobility does not directly follow the size of

DOI: 10.1201/9781003348979-4

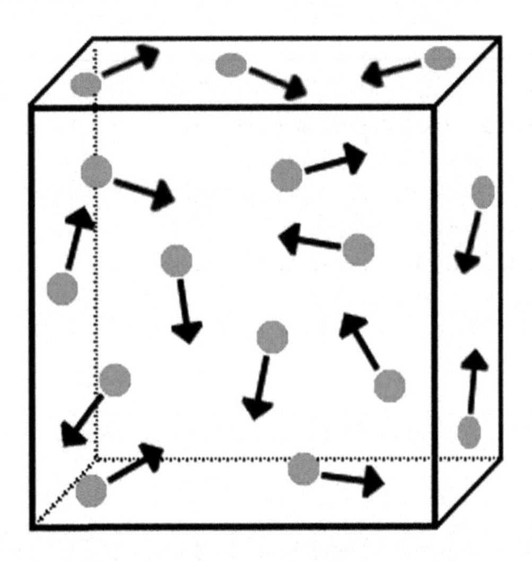

FIGURE 4.1 The arrows depict random movement of particles which are called jumps.

the ion. Presently it seems that these explanations are simple approximations which are ascribing macroscopic properties to a molecular scale.

4.2 IONIC MOVEMENT THROUGH WATER WITH A POTENTIAL DIFFERENCE

As an electrochemical example, in cathodic protection (CP) in seawater, the ions flow through the electrolytic solution at a rate which is determined by the size of the potential difference between the anode and cathode. The potential field is the potential difference against a linear measurement. The electrical field imparts only a slight directional movement imposed on the random motion of the ions. For example, at 1 V/cm this translates to one electric field directed jump to 10^5 random jumps of an ion at room temperature. The positive ions are attracted towards the cathode with the negative ions moving towards the anode. The ionic mobility of these cations and anions is measured practically with a potential difference of 1 V/cm equating to a movement rate of 10^{-4} mm $second^{-1}$ for a typical anion or cation.

One of the reasons why CP is normally successful is explained by the electrical potential measurements shown for a battery in Figure 4.2 where

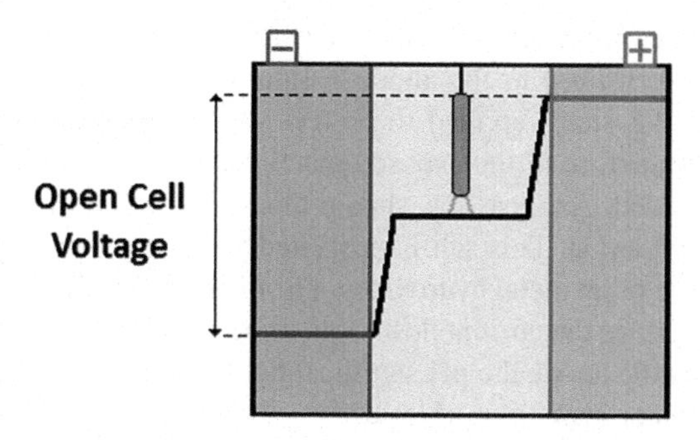

FIGURE 4.2 Ionic flow between two electrodes in a battery showing the change in potential between the cathode and anode in the electrolyte.

there are significant changes in potential near the anode and cathode but for the main part of the electrolyte there is no change in potential. This means that in practical terms there is zero resistance in the main part of the electrolyte. This is because the transport area gets larger away from the electrodes. As V = IR applies the resistance to the electrolytic flow becomes much smaller for the center part of the battery.

The water making up the solution in a battery, CP or corrosion cell is not expected to move significantly (this may be true in certain conditions but not in others as water is a polar molecule and it has been observed moving in certain concrete experiments where a different transport mechanism may predominate as in unsaturated flow described further in Section 2.2 'Flow in Porous Materials'). The rate of movement of the ions is determined by the strength of the electrical field. In early experimentation this was measured in increments of 1 V/cm. If the distance between the cathode and anode is changed from 1 cm to 2 cm with the same driving voltage then the electrical field will be halved so the result is the ions will take four times as long to move double the distance. This linear relationship seems to hold true, at least up to an applied potential of 50 V. Possibly the reason for this experimental finding is the still small number of electric field-induced jumps compared to random jumps even at 50 V.

It is known from practical experimentation that most ions (chloride, sulphate, calcium, sodium, for example) travel at roughly the same speed (at the same electrical field strength) whatever their size. This flow rate does not appear to be related at all to the ions size. There are two significant exceptions to these findings as given in Barrow (1973). These are H^+ which is six times faster than a typical ion and OH^-, which is three times faster. These are probably due to a transfer mechanism where the hydrogen bond migrates rather than the physical masses of the ions.

We would expect hydroxyl ions to be attracted towards the cathode and protons (hydrogen ions) to the anode within a potential field. This is seen to occur on CP systems applied to protect steel in seawater where a whitish film on the surface of the exposed metal cathode is observed after some current has passed. Analysis has shown this to be a film of predominantly magnesium hydroxide. This salt is deposited because of its lower solubility compared to the other metal hydroxides. The result of this film formation is to dramatically reduce the current flow required to provide corrosion protection to the steel. At the anode the pH significantly drops and chlorine gas can be generated in water containing chloride.

Experimental work by Sergi (1995) looked at how the profiles of four ions varied after a circuit was made for a month at an average current density of 350 mA/m^2 and a potential of the cathode of −850 mV with respect to a silver chloride electrode in a simulated concrete pore solution with added sodium chloride. At a 3 V driving voltage the electrical field is roughly 0.4 V/cm. A graph of the data is shown in Figure 4.3.

This shows that there is significant movement of certain but not all ions over the time-period compared to the control specimen, on the right, where no electrical field had been applied. The movement of ions is most pronounced in the 10 mm closest to the anode and the 10 mm closest to the cathode which is similar in profile to the potential changes shown in Figure 4.2. The largest movement is of hydroxyl ions with the halving of its concentration at the anode and increase by approximately 50% at the cathode.

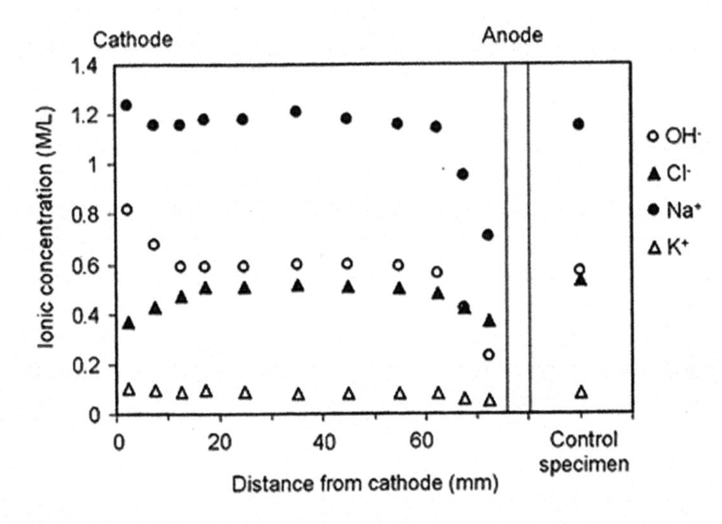

FIGURE 4.3 Profiles of ionic concentrations between an anode and cathode.

That the movement rate of the sodium ions almost matches that of the hydroxyl ions is not what would be predicted with the traditional electrochemical observations which are described earlier in this chapter.

The anions (K^+ and Na^+) also seem to behave very differently, with the sodium being removed from the anode and congregating at the cathode. The potassium ions distribution has not been affected by the imposition of the electrical field. There is also something occurring with the chloride level reducing close to the anode. A simple explanation for this may be chloride gas formation at the anode removing this product from the cell. This is seen practically in CP electrolytic cells where the current densities are substantially lower than in this experiment.

4.3 IONIC MOVEMENT THROUGH CONCRETE WITH A POTENTIAL DIFFERENCE

When concrete sets the hydrated cement consists of solid hydration products plus a saturated water, which is held physically or is adsorbed on the large surface area of the cement hydrates. This is called gel water and is located between the solid products of hydration in gel pores.

The solid products of hydration occupy a smaller volume then the original solid and thus there is residual space in the structure, which are called capillary pores. These can be empty, partially filled or full of water and are about 1,000 nm in width as described by Neville (1996).

The capillary pores are big enough to allow the double layer conduction described in Chapter 1 to occur but the tortuosity of the pore transit paths in concrete will account for a considerably slower transport for the ions compared to brick which is a more porous material than concrete.

When there is potential difference between an anode and a cathode there will be a potential field. In this field the water molecules in the capillary pores will align themselves according to their polarity. This is because the water molecule is polar (not electrically neutral in different parts of the molecule). There is a small net negative charge near the oxygen atom and partial positive charges near the hydrogen atoms. This causes an inner layer of bound water to attach to the pore wall and thus be immobilised. The next layer to this is free to move. This is the so-called electrical double layer. The immobile surface charge surrounding the concrete particle in turn attracts a cloud of free ions of the opposite sign creating a thin (typically 1–10 nm) Debye layer (a layer where there is inter molecular attraction) next to it. The thickness of this electric double layer (EDL) is determined by a balance between the intensity of the thermal random jumps and the strength of the electrostatic attraction to the substrate. In the presence of an external electrical field, the fluid in this charged double layer acquires a momentum, which is then transmitted to further adjacent layers through the effect of viscosity. In this outer layer water can flow taking dissolved salts along with it.

A recent research paper by Schwarz (2022) has been published looking at changes in the ion distribution in a sample with an activated zinc anode using Laser-Induced Breakdown Spectroscopy (LIBS). The use of a sacrificial anode in the experimental specimens limited the driving potential to between 0.3 and 0.6 volt with the anode 25 mm away from the rebar. Thus, the electrical field is roughly 0.2 V/cm. This is a low electrical field level and yet discernible differences in the chloride profile became apparent on samples cut after 7 months, 12 months and 30 months. After 30 months the chloride level has increased from the 0.84% uniform level in the concrete to 13.8% around the zinc anode with a reduction in chloride level around the steel by about 30%. This was compared to a control where the anode was not activated (the anode and cathode were not electrically connected) where no change in chloride distribution occurred.

This LIBS technique was also used to determine the movement of anions namely K^+, Ca^{++} and Na^+. Potassium was used as part of the compound surrounding the anode to keep the anode from passivating so its movements were probably a mixture of current and concentration diffusion. The sodium and calcium ion movements were much clearer with a significant movement of the ions to the cathode (steel) and a significant reduction in concentration around the anode.

4.4 CHANGES TO THE STEEL WITH THE FLOW OF IONS CAUSED BY A POTENTIAL DIFFERENCE

With a potential field and the steel reinforcement as the cathode there will be a transport of hydroxyl ions to this electrode. This will likely cause significant change to the surface conditions of the steel. Some research has been done on this, but this change could be different in the macro and micro corrosion situations, i.e., general corrosion and pitting. Apart from the increase in alkalinity the other large change is likely to be the reduction of chloride by it flowing in the electrical field towards the anode (and possibly by electrostatic repulsion from the cathode) though this could be significantly influenced by the steel reinforcement layout and the location of the chloride contamination.

In the early days of development of CP of concrete there was significant concern that the increase in alkalinity which would be generated by CP could either trigger alkali silica reaction (ASR) or reduce the bond strength. This is the mechanical connection between the steel and the concrete matrix. Limited evidence was found that either of these two factors were significant even at the relatively huge current and voltage levels being applied (250 mA/m^2 at 30 V) but it did introduce the notion that biggest single effect of the CP was to increase the pH of the steel reinforcement generally across its surface and also in the corrosion pits.

This was investigated by Pacheco (2012) in an extended study where some samples had cast in chloride, some had salt applied on the outside and some were cycled in a carbon dioxide-rich environment. The samples were stored outside for ten years. There was clear evidence in some of the samples that active corrosion was occurring after this time. The samples were then cathodically protected and finally destructively examined. From these measurements there was evidence that some parts of the specimens were corroding at a greater rate than others which was a clue about orientation of the steel being important even on laboratory specimens. The CP was applied at a constant voltage of 1.4 V with an initial current density of 30 mA/m^2 dropping to 10 mA/m^2 of steel reinforcement area within 24 hours. The experiment concluded that the acidity in the pits was neutralised within 8 hours of current application and this caused the re-passivation of the steel. The current densities used in these experiments precluded the use of a galvanic anode system.

One of the most important mechanisms explaining how CP works is that as pitting or general corrosion is stopped at more corroding locations the resistance of this ionic pathway increases causing the current to flow to another locations. This effect is commonly seen when a CP system is energised in a relatively dry atmosphere as the resistance rapidly increases in a matter of hours and continues to then incrementally increase until ultimately reaching a stable level at a small fraction of the initial level. This implies that the affect of the current is to improve the barrier properties of the passivation film on all of the steels surface. When in a marine environment, possibly with saturated concrete, often the amount of current required stays relatively stable over an extended period of years. This implies that the passivation film remains incomplete probably due to the transport of oxygen being the rate limiting step.

4.5 CHLORIDE MOVEMENT THROUGH REINFORCED CONCRETE WITH CRACKS

The presence of cracks in concrete structures which are in coastal environments or exposed to de-icing salts may create major durability problems in infrastructure. This statement appears to be obvious and has been researched in many countries. Unfortunately, as with most reinforced concrete research the more you look the trickier it gets as the crack profile as well as width has an effect on its reduction in durability and in real structures you can get crack opening and closing in certain environmental or loading conditions which can also have a drastic effect on durability.

What is known from several studies such as by Lepech (2005) is that there is a threshold crack width which if exceeded can accelerate the corrosion process by orders of magnitude. According to a report by VicRoads (2010) crack widths of 0.3 mm or less do not pose a threat under normal conditions.

When you have wetting and drying in a chloride-rich environment then an enhanced capillary suction will occur pulling the chloride into the structure. It presently appears that the rate of chloride penetration is dependent on the duration of the wetting and drying cycles and how dry the structure gets before wetting again will increase the capillary suction level. This effect is probably hugely important in Arabian Gulf structures as in the night there is commonly fog wetting the structures and then in daylight 50°C temperatures in bright sunshine.

Recently a new technique called LIBS has allowed progressive followings of samples on ongoing experiments in very small areas. In a series of experiments by Savija (2014), specimens were tested with a wet and dry cycle with cracks in profile similar to that seen on a loaded beam. These cracks went from the surface to the rebar and caused some debonding at the steel to concrete interface. It was found that the wider the surface crack the more the chloride penetration occurred. In wide cracks penetration of chloride parallel to the steel occurred due to damage at the steel to concrete interface. Further it was found that the performance of ordinary Portland cement specimens was much worse than slag specimens with more chloride penetration at the top and base of the crack. It was also noted that this behaviour was not replicated in the same specimens without reinforcement. In this situation the cracking was not important in chloride ingress.

REFERENCES

Barrow G, *Physical chemistry*, 3rd edition, 1973, McGraw Hill Publication, p635

Lepech M, Li V, Water permeability of cracked cementitious composites, Proceedings of ICFII, Turin, Italy, 2005, p113–130

Neville A, *Properties of concrete*, Longman, ISBN 0582230705, 1996, p32

Pacheco J et al, Assessment of critical chloride content by EDS revisited, 3rd international conference on concrete repair, rehabilitation and retrofitting, Volume 1, 2012

Savija B, et al., Chloride ingress in cracked concrete; a laser induced breakdown spectroscopy (LIBS) study, *Journal of Advanced Concrete Technology*, 2014, 12, 425–442

Sergi G, Page L, Advances in electrochemical rehabilitation techniques for concrete, UK corrosion, November 1995, 95, 21–23

Schwarz, et al., Ion distribution in concrete overlay, mapped by laser induced breakdown spectroscopy, modified by an embedded zinc anode, ICCRR, 6th international conference on concrete repair, rehabilitation and retrofitting, Cape Town, South Africa, October 2022

VicRoads, Australia, TN038, December 2010, Cracks in concrete

Passivation and depassivation

5

5.1 INTRODUCTION

An engineering metal such as steel is in high-energy state and thermodynamically wants to return to a low-energy state. The time taken to return is primarily decided by the interface between the metal and the surrounding environment. Commonly on the surface of the metal a film of some description will form. The characteristics of this film are extremely important to the corrosion behaviour of the metal. If this film attains a complete covering which is not disrupted the metal loss will be very small and this is called a passive state. For steel if the film is being disrupted directly or oxygen is reduced at one place then iron can pass into solution at another place with the iron oxide appearing at a third place where it does not interfere with continued attack. In this scenario the corrosion rate is significant and this state is known as depassivation.

5.2 FILM FORMATION

On all engineering metals there is a layer on the surface of the metal which is typically an oxide. On steel there is commonly more than one layer which comprises the interface with its environment. This film provides a corrosion-resistant surface layer, the effectiveness of this layer in preserving the mechanical properties of the crystal structure within defines the life of the steel. Many of the properties of the films are controlled or influenced by local defects or non-uniformities in the barrier layer, such as stress-induced breaks in the oxide, local regions of crystalline oxide and, on a larger scale, the effects of second phase particles at the metal surface. The growth of oxide or

DOI: 10.1201/9781003348979-5

other films is accompanied by the development of stresses in the oxide films, which can be both beneficial and harmful to its durability. The ultimate thickness to which an anodic oxide film may grow without cracking or detachment possibly depends on the barriers ability to relieve these growth stresses. As an example of the importance of the nature of the oxide film, titanium which is one of the most reactive metals of all is immune to most corrosion because of the excellence of its oxide layer in reforming and its tenacity.

The presence of stresses in the oxide films and their effects on cracking, spalling and decohesion of oxide films has been recognised for some time and are described by Krishnamurthy (2003). There have been relatively few attempts to investigate the mechanisms that produce the observed stress distributions in films produced by the oxidation of steel in concrete. The earliest explanation for the origin of stresses in a barrier oxide film was offered by Pilling and Bedworth (1923), who proposed that they were due to the difference in molar volumes of the oxide and the metal. Stresses generated within the oxide films may be either compressive or tensile. More recent research by Archibald (1997) has shown that this may be a significant oversimplification as indirect stress measurements have shown that the magnitude and sign of the stresses is dependent upon the oxide formation conditions such as surface condition, pH, current density and electrolyte composition all of which may vary in the concrete corrosion cell.

It is sometimes assumed that the oxide film formed on iron at room temperatures is uniform in thickness across a specimen, and that where several oxides exist together they are present as layers stacked neatly one above the other but this not the case for modern rebar.

With rebar production the process of hot rolling produces mill scale. Mill scale is a mix of metallic iron and three types of iron oxides: wüstite (FeO, grey green in colour), haematite (Fe_2O_3 bright red to dark red in colour) and magnetite (Fe_3O_4, black in colour). It also contains traces of non-ferrous metals, alkaline compounds and oils from the rolling process.

Mill scale is bluish-black in colour. It is usually less than 0.1 mm thick, and initially adheres to the steel surface and protects it from atmospheric corrosion provided there are no breaks occurring in this coating.

Because mill scale is electrochemically cathodic to steel, any break in the coating will cause accelerated corrosion of any steel exposed to the atmosphere. Mill scale provides some protection for a while until the coating breaks due to handling of the steel product or moisture-laden air getting under it. This is why steel fixers used to leave rebar delivered freshly rolled from mills out in the open to allow it to 'weather' until most of the scale fell off due to atmospheric action. This practice is not followed currently and rebar as delivered is cast into concrete on site. Further corrosion of the steel substrate after mill scale formation in exposed atmospheric conditions will tend to have a bright red or orange colour which is indicative of haematite.

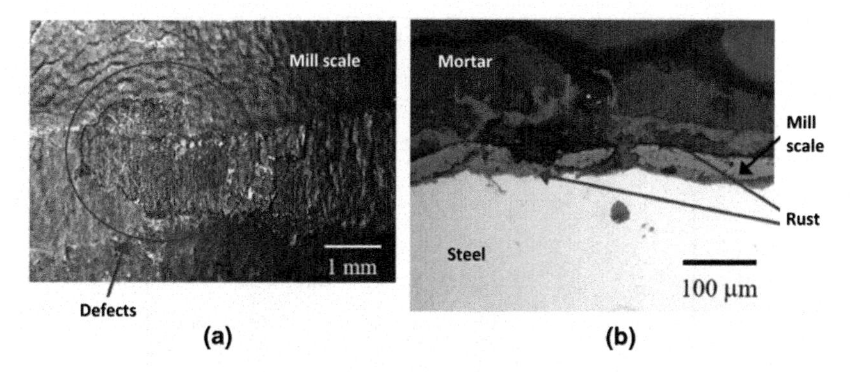

FIGURES 5.1 (a) Mill scale coating on steel showing gaps. (b) Showing mill scale and other corrosion product on a sample in mortar.

After the concrete casting process the mill scale stays inert but other oxides, such as haematite, become converted to give a dull grey colour of passivated steel in concrete. The additional corrosion conversion is shown in a micrograph Figure 5.1.b. This is due to the alkalinity of the pore water initially favouring the anodic formation of a film of a ferrous compound, which is usually ferrous hydroxide, but in the presence of anions yielding less soluble ferrous salts such as carbonates the composition may be mixed. The ferrous compounds are typically subsequently oxidised to an anhydrous oxide which has trivalent iron ions.

There is a continuous movement of atoms and ions at room temperature with factors such as concentration diffusion and electric field potential having important effects on the surrounding film around the steel. As the film increases in thickness, assuming the film is connected to the steel substrate, the distance iron atoms will have travel before they can react, increases and the corrosion rate will diminish. This is diffusion rate control.

If the iron atoms and other ions are meeting and then there is a considerable time for the ionisation process to occur this is activation rate control. As shown in Figure 5.2 the reaction interface has a different profile depending on the reaction control mechanism.

It should be noted that both these control processes are kinetic and not thermodynamic which sometimes gets confused in corrosion process literature. In a recent publication by Googan (2022) it was argued that activation control is the dominant rate limiting step for dissolution of iron in ground water or salt water. This is probably not the case for steel in concrete as the high pH of the setting concrete promotes further film formation as denoted by the colour changes of the oxide film described above.

Steel when exposed directly to an industrial atmosphere (not in concrete) forms a corrosion product of $Fe_2O_3{\cdot}H_2O$ which being loosely adherent, does not form a full protective barrier that completely isolates the metal from the

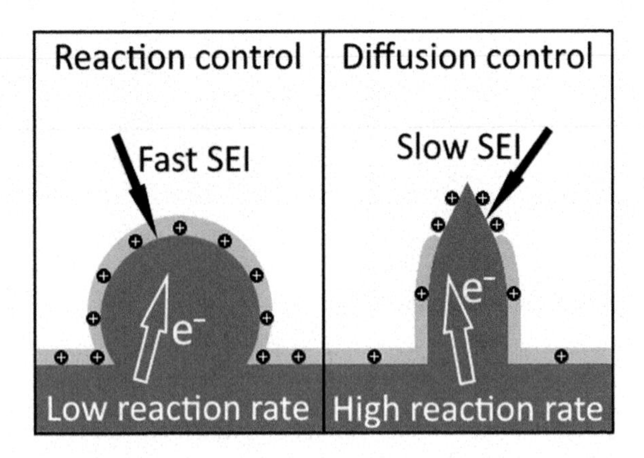

FIGURE 5.2 Drawing of activation and diffusion control mechanisms. SEI is the solid to electrolyte interface.

environment; thus the corrosion reaction proceeds at a slowly reducing rate until all the metal is consumed.

In the case of steel in an alkali solution with no contamination it has been observed that an adherent layer is formed and the corrosion rate measured drops to effectively zero. The alkali simulated pore solutions (SPS) are frequently assumed to behave in the same way as hardened concrete but some research has been undertaken by Duffo (2016) which showed that there was a significant difference in behaviour between a SPS and concrete specimens. It was found first that a protective film formed much more slowly in concrete than in the SPS. It was not clear if it was occurring with same mechanism. Second, the chloride level for triggering depassivation and other factors were significantly different in the SPS compared to concrete. Other studies have looked at how the moisture and oxygen level of the curing concrete will affect the composition of the film. It seems reasonable to suggest that if the film has a different composition it will have different both physical and chemical properties.

For steel in reinforced concrete the red corrosion product on the rebar in addition to the mill scale before it is covered with concrete is probably a version of haematite (Fe_2O_3). The oxide film changes colour to black which is probably magnetite (Fe_3O_4) several months after casting. This magnetite is more adherent and likely to provide a more significant barrier to diffusion than the haematite. A recent study by Alhozaimy (2016) looked at curing environments with varying oxygen levels. They used varying surface conditions (sealed, partially sealed and unsealed in air and submerged in water. It was found that the adequate availability of free oxygen during the initial curing stage was crucial for the proper formation of the passive layer. The study did not look at the long-term consequences of these findings. For example, if the concrete was originally cast under water but was then dried out would it behave in the medium term the same way as a specimen not water cured?

5.3 BREAKDOWN OF THE FILM

It has been shown that the passive films of iron in highly alkaline electrolytes consist of Fe^{2+}-rich inner oxide layers and Fe^{3+}-rich outer oxide layers. This multi-layer film structure is in quantitative agreement with theoretical passivity models that suggest an inner barrier layer forms directly on the metal substrate, and the outer layer precipitating via the hydrolysis of cations ejected from the inner layer.

The big problem in practice is that chloride or similar halide ions in some way interact with this passive film. We know practically that the ions mainly sourced from de-icing chemicals or marine salts, can break down the iron passive film and cause active corrosion. Despite recent advances in nanoscale characterisation of iron passivity, significant gaps exist in our understanding of the dynamic processes that leads to the chloride-induced breakdown of passive films.

There seem to be two thoughts at present for why this depassivation occurs. The first idea is that chlorides penetrate through the passive film by an ion exchange processes or simple diffusion. The second is that the depassivation process is initiated by local acidification of the electrolyte near the film surface, followed by iron dissolution into the electrolyte, and iron vacancy formation in the passive film. In this model the chlorides do not have to penetrate the passive film, but mainly act as a catalyst for the formation of iron vacancies, which diffuse towards the metal/oxide interface, suggesting a depassivation mechanism consistent with the point-defect model. This is where the metal atoms diffuse from the metal matrix outwards leaving a void on the metal surface. Once this void grows to a certain size the process of pitting can begin. These are small localised breakdowns in the film at particular areas where certain unusual conditions exist. A recent study of iron in alkaline media by Dormohammadi (2019) suggests that the point defect model is the more likely corrosion mechanism.

Electrochemical studies with iron in solution have shown that sufficient concentrations of chloride ions are needed at the film/electrolyte interface to initiate the depassivation process and this threshold concentration increases with the electrolyte pH; however, it is not known exactly why this 'critical chloride threshold' exists and why it is pH dependent. The concept of 'induction time', which is the delay in the breakdown of the passive film at chloride concentrations beyond the critical thresholds is also not well understood. As has been discussed previously these SPS studies have only, at best, a limited translation into air exposed concrete probably due to the composition of the films being different but it could also and probably is influenced or controlled by an electrochemical effect.

Chloride-induced iron dissolution and corresponding iron vacancy formation in the outermost (first) layer of the passive film could take place in four stages. In the first and second stages, chlorides facilitate the consumption of hydroxide ions from the electrolyte film on the surface of the iron to form $Fe(OH)_3$ and $Fe(OH)_2Cl$, respectively, which remain stable on the surface of the metal. These two processes cause local acidification, and eventual depletion of OH^-, in the electrolyte adjacent to the surface. The third and fourth stages, lead to the dissolution of iron into the electrolyte in the form of $Fe(OH)Cl_2$ and $FeCl_3$, respectively. These species further dissociate releasing the chlorides back into the electrolyte. Note that both these complexes are trivalent and it has been noted that the presence of chloride favours these ions. The chloride in this is series of reactions is thus a catalyst in the iron dissolution process. There is some evidence for this occurring in concrete with green rust commonly being seen on recently broken out reinforcement as shown by Francois (2018).

REFERENCES

Alhozaimy A, Hussain R, Al-Neyheimish A, Significance of oxygen concentration on the quality of passive film formation for steel reinforced concrete structures during the initial curing of concrete, *Cement and Concrete Composites*, Jan 2016, 65, p171–176

Archibald L, Internal stresses formed by the anodic oxidation of titanium, *Electrochemica Acta*, June 1997, 22(6), p657–659

Dormohammadi H et al, Investigation of chloride induced passivation of iron in alkaline media by reactive force full molecular dynamics, *Materials Degradation*, 2019, 3, 19

Duffo G, Farina S, Electrochemical behaviour of steel in mortar and in simulated pore solutions: analogies and differences, *Cement and Concrete Research*, Oct 2016, 88, p211–216

Francois R, Laurens S, Deby F, *Corrosion and its consequences for reinforced concrete structures*, ISTE Press, ISBN 9781785482342, 2018, p25

Googan K, *Marine corrosion and cathodic protection*, CRC Press, ISBN 9781032105819, 2022

Krishnamurthy R, Srolovitz P, Stress distribution in growing oxide films, *Acta Materiala*, May 2003, 51(8), p271–2190

Pilling N, Bedworth R, The oxidation of metals at high temperatures, *Institute of Metals*, 1923, 29, 529

Electrochemical theory

6

6.1 INTRODUCTION

The tendency of a metal to oxidise can be related to its electrode potential in an idealised situation. The equilibrium electrode potential of a metal corresponds to the equilibrium of its oxidised and reduced species. At the equilibrium potential the metal and its ions coexist. When the potential of the electrode is changed in one direction the ions will not be stable and thus corrosion occurs. In the other direction the metal will be stable and this is commonly referred to as immunity. This reversible process and the energy associated with it are the bedrock of electrochemical principles and are discussed further below.

6.2 REVERSIBLE POTENTIAL AND THE SINGLE ELECTRODE

A single electrode is the interface between a metal and its ions in an aqueous solution across which a charge transfer results from an electrochemical reaction. When the number of metal atoms being oxidised to ions and ions reduced to metal is equalised an electrochemical equilibrium will exist and this potential can be measured using a calibrated reference electrode. Typically in the laboratory a standard hydrogen electrode (SHE) is used. This value has been recorded for most of the metals in water and allows a scale of increased ionisation to be constructed. Thus, a corrosion tendency as the metals move from behaving in a noble manner to becoming active can be assembled. Gold is at one end of this scale and lithium at the other.

DOI: 10.1201/9781003348979-6

6.3 ELECTROCHEMICAL CELLS

The immersion of two metal plates in an electrolyte constitutes an electrochemical cell. The two plates are connected by either a wire or an electric power source between them. If the two metal plates are of different metals there will be a current flow through the wire with the more active metal acting as the anode and the more noble metal acting as the cathode. This is a galvanic cell. The anode is undergoing an oxidation process and the cathode a reduction reaction. An electrolytic cell is where current is forced by an external source and common examples of this are impressed current cathodic protection and electroplating of metals.

6.4 REVERSIBLE ELECTRODES AND THE NERNST EQUATION

In certain situations you can have a reversible electrode. For example, this can be thought of as a lead acid car battery. When the voltage is over a certain level the lead ions will coat the battery plates with lead metal and below a certain voltage the lead metal will go into solution giving electrons and thus power. Incidentally and unfortunately this reversible electrode does not occur with iron as you get the iron going to iron ions and electrons but reversing the electrical current produces hydrogen gas. Thus, iron lost in the corrosion process cannot be made good by an electrochemical process. For this reason iron is not used in most textbooks to illustrate this process.

An ideal reversible electrode (where either anode or cathode reactions can occur) has no resistance and thus the energy measured would be the difference between the state of reduction and oxidation. All real electrodes have a finite resistance to charge transfer at the metal/solution interface. This is commonly known as Faradiac resistance and is caused by a double-layer capacitor alignment forming on the electrode surface. This capacitance derives from separation of electrical charges which is generated by the directional alignment of ions and electrons at the interface between the electrode material and electrolyte.

At very low current densities there may be little change in the double-layer composition so roughly the energy involved in the anodic and cathodic processes is the energy involved in the difference between the reactants and products. This greatly simplifies things for calculations and so in these conditions where the reaction proceeds at a very low rate at the anode it is reasonable for certain reactions to remove this double-layer capacitance from the calculations as shown pictorially in Figure 6.1.b. With this factor removed and using Faraday's first law of electrolysis (the chemical deposition due to the flow of current through an electrolyte is directly proportional to the quantity of electricity (coulombs) passed through it) then the relative activities of the metal against its ions can be presented. Normally in electrochemical parlance activity is substituted for concentration as this is much easier to determine. Faraday's law is that the electrons that transport electricity will react with a definite quantity of the metal to produce ions or with a different polarity, ions to metal. The electrochemical equivalent is the molar mass of the substance deposited to one of the electrodes when a current of 1 A is passed for 1 second. This amount of electricity is defined as 1 C (Figure 6.2).

The Nernst equation provides a value of the reversible potential as a function of the ionic concentration in the solution. As an example for Cu^{++} at 1 g ion/L of electrolyte the reversible potential is +340 mV while Cu^{++} at 10^{-6} g ion/L of electrolyte the reversible potential is +160 mV relative to a SHE. All the standard electromotive potentials are measured at this 1 g of ion/L concentration (Figure 6.3).

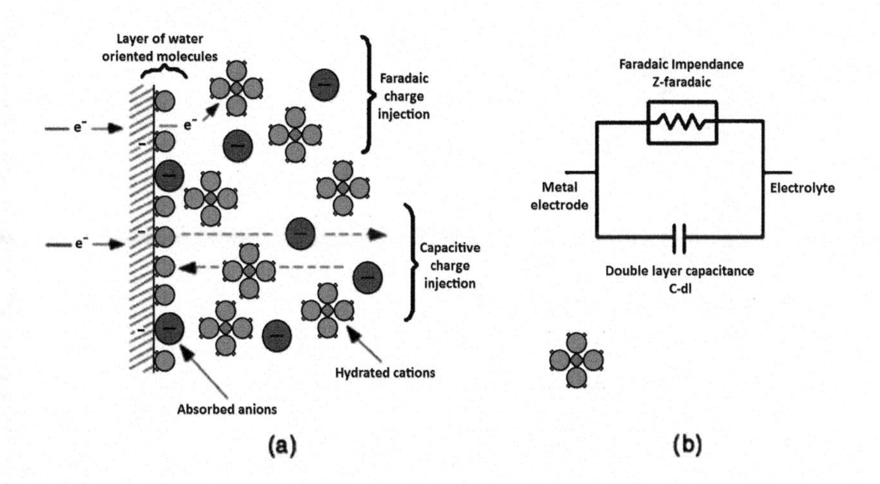

FIGURE 6.1 Double layer capacitance on an electrode. Pictorially shown in (a) and as an electrical schematic in (b).

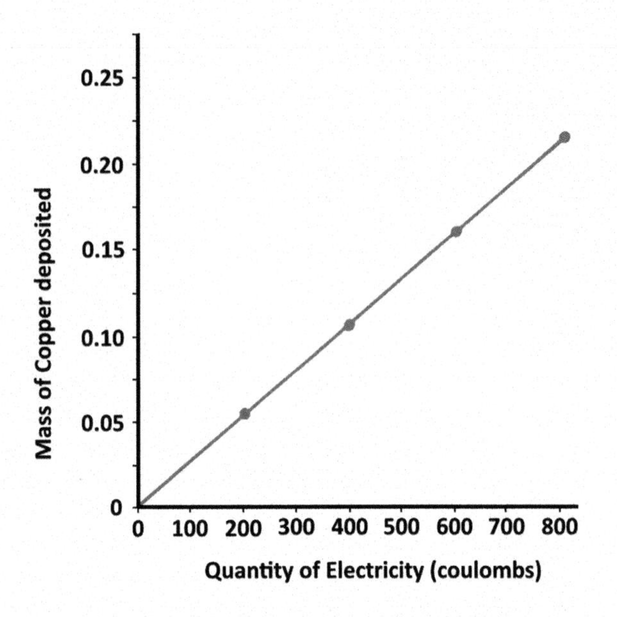

FIGURE 6.2 Faradays first law of electrolysis showing good correlation for an experiment where copper metal deposition is plotted versus current passed.

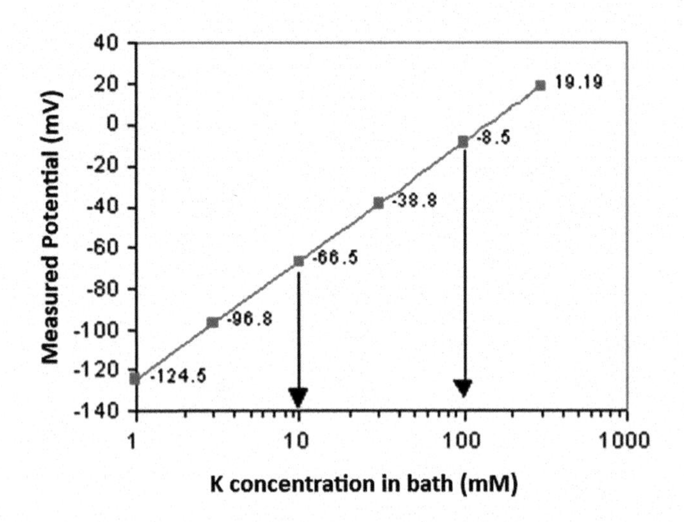

FIGURE 6.3 Practical display of a Nernst logarithmic relationship between concentration and measured potential.

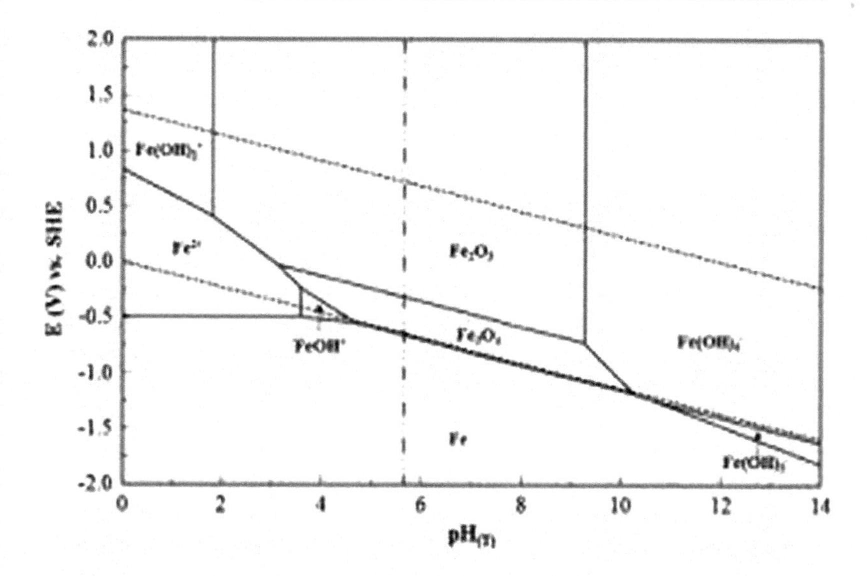

FIGURE 6.4 Pourbaix diagram for iron in water.

6.5 THE POTENTIAL PH DIAGRAM

A further addition to the above reversible potential for an element is to introduce pH or concentration of hydrogen ions. These are commonly referred to as Pourbaix diagrams after their inventor Marcel Pourbaix. An example is given below for the iron species against pH (Figure 6.4).

The potential -pH diagram is a tool to predict the chemical reaction likely to proceed in a particular environment and so it can provide information on what stable ion species will be ultimately obtained.

6.6 TAFEL PLOT AND EVANS DIAGRAM

Julius Tafel studied the Hydrogen Evolution Reaction (HER) in early 1900. This commonly is the splitting of water using electricity. He found that there is an exponential relationship between the applied current at a platinum surface in water and the potential.

This is also true the other way around (applied potential and measured current). A convenient way of plotting this relationship was to plot the potential (E) versus the logarithm of the current, log I, because using the logarithm leads to a linear plot. Why logarithmic and not some other relationship? I have looked hard at this and the only answer I can come up with is this what early electrochemists found and it is also commonly seen in other chemical phenomena such as the pH scale and the Nernst relationship.

In Figure 6.5 the slope of the line is called the Tafel slope. It is usually expressed in the units mV/decade. A decade is to the power of ten as the x axis is on a logarithmic scale. This example is an idealised case and thus is not commonly seen in a practical experiment. The reason for this, for example in the oxygen reduction reaction (ORR).

$$O_2 + 4e^- + 4H^+ = 2H_2O$$

where oxygen gas is reduced to water, is that almost immediately the oxygen is depleted within reach of the electrode. The reaction can only continue, and thus a current can only occur, if new oxygen diffuses towards the electrode. The current then no longer depends on the potential, but the transport of oxygen in the solution to the electrode. So the Tafel plot will no longer be linear as is shown in Figure 6.6.

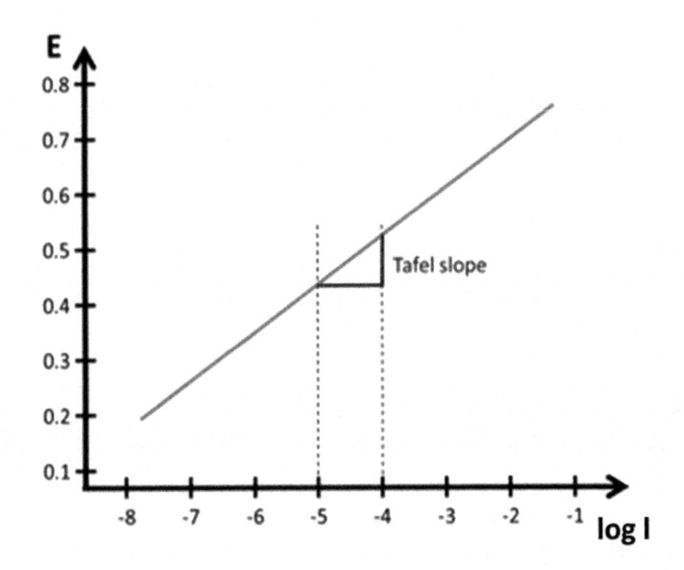

FIGURE 6.5 Tafel plot scheme with arbitrary scale and indication of the Tafel slope.

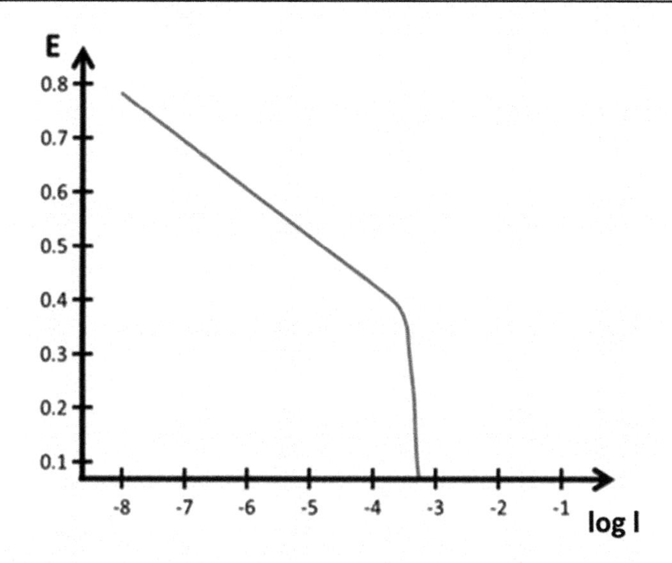

FIGURE 6.6 Tafel plot of a diffusion limited system.

6.7 COMBINING REDUCTION AND OXIDATION IN AN EVANS DIAGRAM

There is a simultaneous reduction and also an oxidation for corrosion to occur, i.e., there must be both an anodic and cathodic reaction at the same time. In reality the anode and cathode can be significantly physically separated (for example, in a modern pipeline cathodic protection the separation can be the order of a hundred miles), but here it is on the same electrode and thus these are known polyelectrodes or mixed electrodes. For copper there is metal going into ions and vice versa and thus you can have an equilibrium potential as given by the Nernst equation. However, in the example of iron immersed in an acid solution, with hydrogen ions, the cathodic process is hydrogen gas evolution and the anodic process is the reduction of iron that both proceed irreversibly. The potential that is measured is a mixed potential and because the iron being oxidised as an active process you are actually measuring the corrosion potential known as E_{corr}.

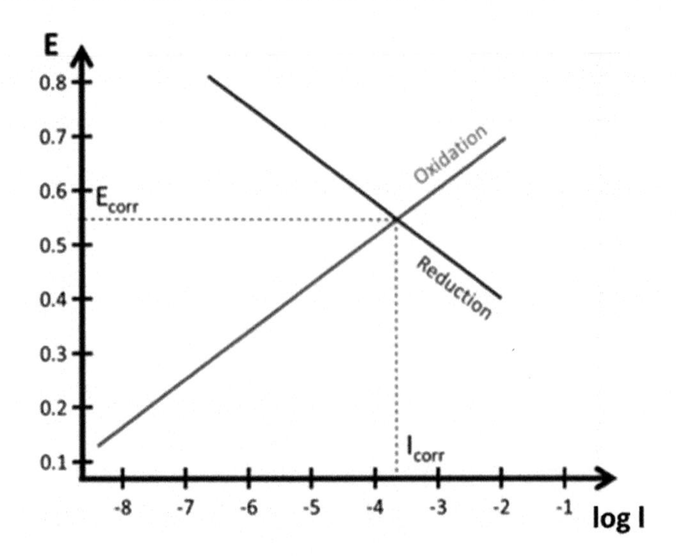

FIGURE 6.7 Evan's Diagram showing the two Tafel plots of the oxidation and reduction reactions.

If the Tafel plot of both the reduction and the oxidation reactions is known, the two Tafel plots can be used to find the theoretical corrosion current and corrosion potential. This is possible due to two facts:

1. An immersed conducting sample has one potential at any instant and thus all reactions must happen at that potential.
2. The conversion of charge demands that all electrons donated need to be accepted, i.e., the reactions have to happen at the same rate which implies the same current.

From these two conditions it can be derived that the corrosion current and the corrosion potential are determined by the point where the two Tafel plots of the reduction reaction and oxidation reaction meet. Plotting the two Tafel slopes into one plot is what is done in an Evans diagram as shown in Figure 6.7. This is helpful to estimate what influence a change in the oxidation or reduction rate has on the corrosion rate. Also the potential and corrosion current of a galvanic couple can be predicted.

6.8 POLARISATION CURVES

The Evan's diagram can practically only be used for qualitative estimations. The missing quantitative data usually makes it necessary to evaluate the system with an experiment using a potentiostat or galvanostat. These apparatus changes the potential (typically 1 mV/second but it is practically determined in a laboratory to be as quick as possible while not having a drastic effect on the results) in a sweep on both sides of E_{corr} with the current recorded by the apparatus electronics. For example, an analogue solution with calcium hydroxide and sodium chloride to imitate pore water and a cleaned polyanode the sweep is typically 10 mV on either side of E_{corr}.

The recorded current is the difference between the current of the oxidation and the reduction reactions. This means that the measured current at the corrosion potential is 0. Since the plot is made in a logarithmic scale a 0 would correspond to a minus infinite (−∞), which a potentiostat cannot measure. A schematic of a polarisation curve is shown in Figure 6.8.

The goal of recording a polarisation curve is usually to extract the corrosion potential as well as the corrosion currents, but as in the previous paragraph discussed the point of interest, the intersection of the two Tafel plots, is not directly visible in the polarisation curve.

Further away from the corrosion potential the polarisation curve is mainly influenced by only one of the reaction. At very cathodic potentials the reduction dominates and at very anodic potentials the oxidation. Due to this the linear parts of the polarisation curves are used for the calculation of the Tafel slopes and thus the corrosion potentials as well as corrosion current.

For a reliable extrapolation the more linear the behaviour the better. Again the more decades show this linear behaviour the better the reliability of the extrapolation.

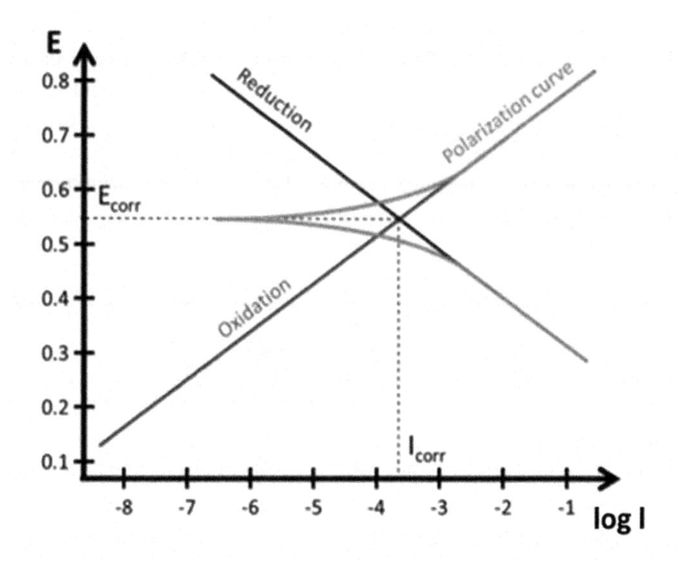

FIGURE 6.8 Polarisation curves with Evan's diagram (straight lines).

6.9 LINEAR POLARISATION RESISTANCE

Linear polarisation is a non-destructive method for measuring polarisation resistance of steel in concrete. Ideally it could provide an indication of corrosion rate at or near the free corrosion potential of the reinforcement under the measuring head. It works using a reference electrode (called the working electrode) to measure the free corrosion of the reinforcement. Subsequently a counter electrode in the head which is in contact with the concrete surface applies current in a cathodic direction followed by an anodic direction with the working electrode measuring the change in potential of the reinforcement. The change in potential is governed by the polarisation resistance. From this a corrosion rate per year can be calculated using a formula derived from Faradays law with a constant B set at either 26 or 52 mV depending on the passivity or active condition of the steel.

6.10 ACOUSTIC EMISSION

When metal corrodes there is an emission of sound and this monitoring technique is widely used on vulnerable tendons on steel suspension bridges as an early indicator of corrosion problems. It has also been attempted on reinforced concrete structures but this complicates the matter. A recent review of the current position relative to monitoring reinforced concrete was undertaken by Verstrynge (2022) who also looked at developing the mainly laboratory-based protocols into accelerated corrosion testing. The future of on-site acoustic emission assessments of corroded reinforcement was also considered. At best this a technology of the future at worst it is expensive and unlikely to be definitive.

6.11 CONCLUSIONS

For almost 200 years (1831 Faraday's law) some aspects of electrochemistry have been understood and attempts have been made to rationalise what is being recorded in laboratory corrosion experiments and occasionally from structures back to this and later equations. This initially looks fairly reasonable position to follow. While these broad principles still hold true they are a simplified overview of what now appears, in the case of corrosion of steel in concrete, to be an extremely complex series of chemical reactions intertwined with electrochemical considerations. Reviewing recent text books and research papers there does not yet seem to be a more satisfactory way to categorise these reactions other than by using principles developed in the 1930s.

REFERENCE

Verstrynge E et al, Steel corrosion damage monitoring in reinforced concrete structures with the acoustic emission technique: A review, *Construction and Building Materials*, September 2022, 349

Electrochemistry reality of steel in concrete

7

7.1 INTRODUCTION

Corrosion occurs because the metal is in a higher energy state than the oxide and thus there is a thermodynamic driving force for this reaction. Thus the important question is kinetic and how quickly this will occur. This is directly affected by the metal's environment. For an extreme example, steel in hydrochloric acid can be significantly attacked in minutes as demonstrated in a paper by Noor (2008) whereas in a dry air environment iron meteorites can remain unchanged for millions of years. This denotes a reaction rate difference of trillions of times.

It has long been known that significant corrosion to steel reinforcement in reinforced concrete structures occurs in certain exposure environments and more latterly it has been noted that particular designs of structures are vulnerable.

Due to the massive practical importance of this particular corrosion reaction significant efforts in many countries over the last 40 years have been made to define this corrosion reaction using standard electrochemical research practice which is discussed in Chapter 6 and by some other non-destructive methods. The corrosion parameters such as corrosion potential, corrosion current density with the anodic and cathodic Tafel constants have been recorded by various experimenters. These parameters are required to corrosion model this reaction in accordance with electrochemical principles discussed previously. The results of some of the various experiments by researchers are given below.

DOI: 10.1201/9781003348979-7

7.2 TAFEL SLOPES MEASURED OF STEEL IN SIMULATED PORE SOLUTIONS (SPS)

Various researchers have undertaken experiments in pore solution-type water with chloride added to determine when rebar samples make the transition from active corrosion to passive state and vice versa. A slightly different technique called linear polar resistance (LPR) has also been used. The table below is an attempt to show the diversity of the values recorded by the different researchers

RESEARCHER	*ACTIVE CORROSION (mV/DECADE)*	*PASSIVE CONDITION (mV/DECADE)*
LPR (Stern-Geary equation)#	26	52
LPR (Stern-Geary equation)*	26	52
Sohail (2021)	54	47
Warkus (2006)	100	Infinity
Redaelli (2006)	75	1,000
Ge (2007)	60	60
Kim (2008)	60	60
Duprat (2019)	637	218

As measured by Andrade and Gonzales (1978)
* As calculated by Sohail (2021)

As can be seen there are very significant reported differences between the various researchers in what might be thought of as a fairly reproducible experiment. Duprat (2019) is reporting that the slope significantly increases in active corrosion. This is contrary to all the other researcher's results and also contrary to the accepted wisdom of passivation against active corrosion.

7.3 TAFEL SLOPES MEASURED OF STEEL IN CONCRETE

Various researchers have undertaken also undertaken experiments in concrete (C) and chloride contaminated concrete (CC) to determine when rebar samples show the transition from active corrosion to passive state. The table below is an attempt to show the diversity of the values recorded by different researchers

RESEARCHER	*TYPE*	*ACTIVE CORROSION (mV/DECADE)*	*PASSIVE CONDITION (mV/DECADE)*
Babaee (2016)	CC	430 to infinity	106–221
Babaee (2016)	C	Infinity	30–45
Michel (2016)	CC	10–369	10–233
Garces (2005)	CC	73–242	

As might be expected there is more variation in the values recorded in concrete then in the SPS. The active corrosion reported ranges from 10 mV per decade to infinity mV per decade. There is a huge range in values recorded by each of the researchers who presumably used a homogenous approach. In order of researcher; an infinite difference, a factor of 36 and a factor of 3, respectively. With this level of uncertainty it is very difficult to think that these values have any practical purpose let alone feed into the Butler-Volmer equation which is a series of equations to define a uniform corrosion system under an imposed polarisation. The Tafel slopes above for active corrosion (B_a) provide a level of qualitative representation of the anodic curve on the Fe/Fe^{++} electrode. A low-coefficient value reflects (high mV per decade) the behaviour of active steel with a high-coefficient value the behaviour of passivated steel.

7.4 CORROSION RATES RECORDED IN ACTIVE AND PASSIVE STEEL IN REINFORCED CONCRETE

With experiments on concrete specimens in a laboratory the following was measured

RESEARCHER	*ACTIVE CORROSION (mA/m^2)*	*PASSIVE CONDITION (mA/m^2)*
Sohail (2016)	3–26	0.1–1.5

The units have been converted from current per cm^2 as reported to current per m^2 as this unit is used in practical corrosion engineering associated with reinforcement corrosion. The values recorded in this experiment accord with both the values of current applied in cathodic protection installations for steel in concrete. Also they are in alignment with the definition of passivity in EN12696 which specifies a corrosion level of less than 1 mA/m^2 of steel surface area. These values are in good alignment with the cathodic protection levels for remediation in a range of 2–20 mA/m^2 and in new (hopefully passive) concrete of 0.2–2 mA/m^2 in the above standard. The values reflect an earlier British standard BS 7361:1991 which gave values of 5–20 mA/m^2 for remediation which was obtained from data on working reinforced concrete cathodic protection systems.

7.5 PASSIVE AND ACTIVE CORROSION POTENTIALS MEASURED IN REINFORCED CONCRETE STRUCTURES

The potential of the steel measured from the surface of the concrete has proven to be an effective guide to the likelihood of active corrosion occurring. It is now commonly used as a survey technique and has been verified on many occasions by destructive removal of the cover.

Steel in concrete has a potential which is a balance of the anode reaction and cathode reaction potentials. Extremes in potential are thus recorded when either the anodic or the cathodic reaction predominates.

For example in dry, high-quality concrete the steel behaves like a noble metal with no corrosion with the potential measured being the cathode reaction only. This potential is the reaction of oxygen to form hydroxide which is approximately +180 mV with respect to a silver/silver chloride/potassium chloride electrode. In a water saturated, oxygen-free environment the potential corresponds to the anode reaction. This is the Fe to Fe^{++} oxidation reaction which is about –910 mV with respect to a silver/silver chloride/potassium chloride electrode.

In exposed concrete the oxygen transport rate can be sufficiently high as discussed by Chess (1996) to allow an average of 300 grams of steel per year to corrode over a surface area of one square meter. This equates to a corrosion current of 30 mA/m^2. This is a similar value for active corrosion to that recorded by Sohail in Section 7.4. The general corrosion profile related to potential is given below (Figure 7.1).

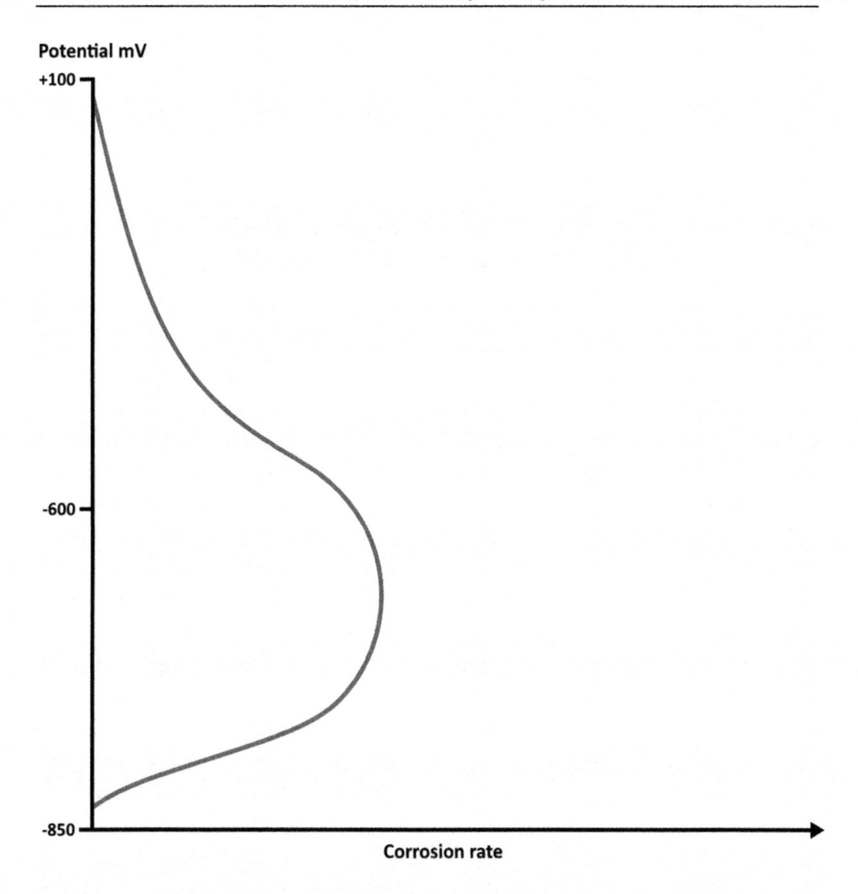

FIGURE 7.1 Relationship between the corrosion rate and the potential of steel against a silver chloride reference concrete at 20°C from Chess (1996).

The above findings are for general corrosion. When pitting corrosion occurs there can be significantly more positive potentials recorded while corrosion is still occurring. A common location for pitting is when there is a disruption to the concrete cover, such as at leaking casting joints. At these locations chloride-enriched water can directly interact with the steel surface causing a localised anode and cathode with a significant potential difference. This favours the pitting reaction and the free water can also allow the transport of the iron ions away from the reaction site preventing any protection from film formation. In this case this corrosion will be significantly increased by the difference in the small anode size compared to the cathode sink of the bulk of the reinforcement. In real structures you can also have both pitting and general corrosion occurring simultaneously either in different or even the same areas.

7.6 MODELLING FOR LIFE EXPECTANCY AND REALITY

A model for predicting chloride ingress into concrete aims at predicting a chloride profile typically at a depth of the first level of reinforcement. Initially a single chloride ingress model was proposed but as more data was collected it was realised that the actual exposure condition of the structure was of great importance The various exposure conditions can be categorised as submerged (foundations), marine (jetties, marine bridges), wicking (tunnels, caissons and swimming pools) and de-icing (land bridges, car parks). These categories all tend to behave in a similar way from a corrosion viewpoint though temperature, mechanical movement among other factors also have a significant effect. These categories are now being modelled separately. Probably the most definitive description of this modelling is given by Luping (2012).

Early chloride penetration modelling was undertaken primarily in accordance with Fick's first law. This states that the rate of diffusion is proportional to both the surface area and concentration difference and is inversely proportional to the thickness of the membrane. An additional factor which is modelled is the chloride binding of the concrete matrix impeding its flow through the material. This model correlates well with actual data for submerged structures but after the first few years of exposure the chloride binding level changes which requires a small modification to the equations. The only required input for this modelling is the apparent diffusivity.

In structures with occasional wetting the above model is adjusted to allow for moisture flow as this wetting and drying can play a dominant role in chloride transmission. These models require a large number of additional parameters to be defined into a mathematical relationship.

The most recent modelling technique describes the chemical interactions between all the ions in the pore solution and matrix. This is normally called the Nernst-Plank model as the fluxes of each individual ion species are predicted and then an estimation of the overall electrical field is made. This is then related to how it will influence the individual ions by making them speed up or slow down depending on their respective polarities. This model is attractive in that it is taking to account the ionic movement which was described in Section 4.3 and is likely to be extremely significant in a true structure. The problem with this approach is that this type of model requires a huge amount of input data, which are not readily available even in a laboratory or simulation situation. It was concluded by Luping that at present most of these Nernst-Planck models will remain as research tools and not be used in practical applications.

A significant effort has been made to evaluate the models against real-life data taken up to 30 years from different categories of structures such as road bridges and marine structures. It was found that a reasonable correlation existed between the models and the structures with a few exceptions. Examples of this are where there was an unexpectedly poor performance of silica fume concrete in bridges and leaching effects in submerged concrete reduced chloride penetration.

There is a real problem for life expectancy modelling and this is not discussed in the comprehensive tome by Lupin. As presented in Section 3.4 'Review of corrosion threshold level' you can have active corrosion at 0.1% chloride level adjacent to the steel and no corrosion at 6%. This means that there are factors which are more critical to the occurrence of corrosion of steel in concrete then the chloride level. There are probably several reasons for these findings with most likely being the electrochemical nature of the corrosion process. If there is no significant potential difference in adjacent areas of reinforcement, no anodes and cathodes will be formed, there will be limited electrolytic movement and no concentration of ions which could trigger the depassivation of the overlaying film and allow active corrosion.

7.7 NON-DESTRUCTIVE TESTING FOR CORROSION OF STEEL IN CONCRETE

For practical structures there are two electrochemical techniques commonly used namely potential mapping and linear polarisation resistance measurement which with three-dimensional nature of the steel below and the heterogeneity of the concrete at best allow a quantitative assessment of the steel corrosion state. Other approaches have been assessed such as acoustic emission which will be extremely expensive, slow and difficult to interpret. Virtually every major conference a new NDT system for use in remotely evaluating the corrosion status of reinforcement is described. Unfortunately, the approaches recently proposed seem to be getting further away from a simple reliable technique.

In contrast a recent approach by Francois (2018) has described in detail the quantification of the corrosion rate using the cracks created by reinforcement corrosion. The approach here is that the pressure exerted by the corrosion product relates to the local cross section loss of steel. This is an empirical approach but seems to be fairly reproducible with a cross sectional loss of 2.6% being sufficient to initiate cracking. This corrosion cracking can be separated from structural cracking as it runs parallel to the rebar as opposed to transversely. This approach is quick, cheap and probably significantly more reliable than all the electrochemical methods apart from potential mapping. Initial cracking may be caused by localised corrosion. This crack may short circuit the barrier effects of the concrete and interface, so it is likely this may change to general corrosion further in the corrosion evolution. Following the progress of the cracking allows this sort of change in the corrosion process to be incorporated.

Francois further looked at the effect of corrosion on the structural properties of elements in certain structures. This is outside the scope of this book though it may be of interest that Rostam (1985) was undertaking in the early 1980s full-scale destructive testing of structures with and without corrosion damage. The results were mixed in that it demonstrated that some civil engineering design premises were incorrect. These researchers concluded the effect of corrosion was fairly minor for columns but significant for beams.

7.8 CONCLUSIONS

As has been apparent for decades corrosion of steel in concrete is an electrochemical reaction. The main trigger for this corrosion occurring is chloride ions. Again this has been obvious for decades as the reinforced concrete structures in trouble in the 1980s were either marine, had de-icing salts or had chloride added as a set accelerator. This last category is now rarely seen as most of the affected structures have been replaced. As the problem is so widespread and expensive to fix there has been ongoing and significant funds employed in multiple countries to research the reasons for this corrosion. One of the approaches adopted is an attempt to define the reaction in accordance with the classic electrochemical principles such as Tafel slopes, linear polarisation resistance, active and passive corrosion rates as described in this chapter. The variations in the experimental data recorded is remarkable and the simplifying assumptions in these laboratory experiments, mean that at best, the information is indicative and does not take us any further forward. When electrically testing concrete specimens the first thing that can be noted is that concrete acts as a capacitor, indeed there is presently active research on concrete derivatives acting as a power storage device. This significant capacitance takes its behaviour away from behaving in a way that the classic electrochemical principles describe.

Presently linear polarisation is commonly used for corrosion assessment with commercially available systems for undertaking this mapping. After decades of research and development the best you can hope for is an indicative result because of the difficulties in defining the steel-reinforcement target. With all the effort into electrochemical assessment it is a little disheartening that a recent publication by Francois (2018) concluded that best method for evaluating corrosion onset and progress was surface cracking.

Useful non-destructive testing on real structures using electrochemical or other properties is presently limited to two versions of potential mapping which can give useful data. The normal measurement is of the steel reinforcement measured against a surface electrode but you can also use reference electrode against reference electrode. In this technique you are looking for potential field gradients. This technique has the advantage of not requiring a connection to the steel reinforcement.

REFERENCES

Andrade C, Gonzales J, Quantitative measurements of corrosion rate of reinforcing steels embedded in concrete using polarisation resistance measurements, *Materials and Corrosion*, 1978, 29, 515–519

Babaee M, Castel A, Chloride induced corrosion of reinforcement in low calcium fly-ash based geopolymer concrete, *Cement and Concrete Research*, 2016, 88, 96

Chess P, Gronvold F, *Corrosion investigation; a guide to half cell mapping*, Thomas Telford Publishing, ISBN 0727725041, 1996

Duprat F, Larrad T, Vu N, Quantification of Tafel coefficients according to passive/active state of steel with carbonation induced corrosion in concrete, *Materials and Corrosion*, 2019, 70, 1934

Francois R, Laurens S, Deby F, *Corrosion and its consequences for reinforced concrete structures*, ISTE Press, ISBN 9781 78548 234 2, 2018

Garces P, Andrade M, Saez A, Alonso M, Corrosion of reinforced concrete in neutral and acid solutions simulating the electrolytic environments in the micropores of concrete in the propagation period, *Corrosion Science*, 2005, 47, 289

Ge J, Isgor B, Effects of Tafel slope, exchange current density and electrode potential on the corrosion of steel in concrete, *Materials Corrosion*, 2007, 58, 573

Gjorv O, *Durability designs of concrete structures in severe environments*, CRC Press, ISBN 97814665 87298, 2014, p30–31

ISO/FDIS 12696, Cathodic protection of steel in concrete, 2022, p39

Kim C-Y, Kim J-K, Numerical analysis of localized steel corrosion in concrete, *Construction Building Materials*, 2008, 22, 1129

Luping T, Nilsson L, Basheer P, *Resistance of concrete to chloride ingress*, Spon Press, ISBN 9780415486149, 2012

Michel A, Otieno M, Stang H, Geiker M, Propagation of steel corrosion in concrete: experimental and numerical investigations, *Cement Concrete Composites*, 2016, 70, 171

Noor A, Al-Moubaraki A, Corrosion behaviour of mild steel in hydrochloric acid solutions, *International Journal of Electrochemical Sciences*, 2008, 3, 806–818

Redaelli E et al, FEM models for the propagation period of chloride induced reinforcement corrosion, *Materials and Corrosion*, 2006, 57, 628

Rostam S, Sorensen K, Bergholt K, Full scale load tests of damaged concrete structures, Deterioration and repair of reinforced concrete in the Arabian gulf, 26–29 October 1985, p333–350

Sohail M et al, Electrochemical corrosion parameters for active and passive reinforcing steel in carbonated and sound concrete, *Materials and Corrosion*, 2021, 72(12), 1854–1871

Warkus J, Brem M, Raupach M, *Materials and Corrosion*, 2006, 57, 636.

State of the art

8

The premature corrosion of rebar in concrete is a very significant problem worldwide and is now argued by Marcus (2020) to be the single most important world degradation problem. It could be argued that plastic in the oceans is worse because it is so difficult to remedy but probably by direct financial cost corrosion of steel in concrete is more of a problem. It should be noted here that steel structures directly in the environment (such as bridges or ships, for example, have easily achieved their design life in aggressive chloride containing environments because regular anti-corrosion maintenance such as painting and cathodic protection is accepted). Because of the wide spread and serious nature of this steel in concrete corrosion problem, in most countries of the world there has been a multitude of laboratory research programs with the purpose of developing improved concrete ingredients which in laboratory testing seem to offer significant advantages in durability. Most of the amendments either increase the tortuosity of the concrete matrix or fill in the pores. On real structures the jury is out on many of the innovations incorporated into concrete over the last two decades, particularly on major structures with very high-strength concretes looking vulnerable where there is dynamic loading (Figure 8.1). Concern has also been raised about the change in composition of the steel in the last few decades but possibly the difference in corrosion resistance will be minor and masked by other much larger variables. To date little has been done to the design of structures beyond increasing cover depths and increasing diffusion resistance of the concrete but there are some recent projects where this has been done, particularly in the Middle East where concrete problems abound. In these designs a more determined effort to prevent wicking and also an effort to reduce the likelihood of anodic areas forming in vulnerable locations has been made. It is worth emphasising that in other countries a remarkably small effort is being made to improve the durability through improving the design of reinforced-concrete structures beyond changing the concrete mix and increasing cover depths.

There has been a significant effort particularly in laboratory conditions to assess if and how much corrosion is occurring. Some of this has been a wasted effort as much of the work has been in simulated pore solutions where

DOI: 10.1201/9781003348979-8

FIGURE 8.1 On structures such as bridges high-strength concrete with dynamic loading has led to significant cracking causing premature corrosion.

the corrosion products and mechanisms are not the same as in a concrete structure and thus, at best, are indicative. From the various research programs 16 factors that in the same conditions will have an effect on the critical chloride content to initiate corrosion has been compiled by Angst (2009). In combination these factors may explain why some real structures are corroding at a low percentage of their design life while others are surviving while both being in an aggressive environment.

Recently new laboratory equipment has allowed a much more direct study of the corrosion process over a time period with reinforced-concrete specimens. The first results have been published and make interesting reading. The machines used are called LA-ICP-MS or LIBS which are probably the same technique with different acronyms. This has changed the game as now the use of concrete samples can be substituted for simulated pore solutions and an analysis of much smaller areas is possible over a substantial time period with voltage gradients included between an anode and a cathode. The equipment to undertake this research is expensive so it will favour the wealthier and larger research departments and even then this testing is still going to have limitations. This is because of the simplicity of the test specimens required for this technique contrasting with the relatively massive three-dimensional nature of the electrical potential layout of a reinforcement cage in a real structure. What the early users of these techniques have shown is that there is a significant to dramatic movement of chlorides under small electric fields and this hints at the main reason why such different chloride levels can either initiate corrosion or not in real structures.

A significant effort has been made on modelling chloride ingress over many years and effective formulas for various categories of environments such as wicking, underwater and other environments have been developed and verified with data obtained from real structures to a reasonable level of reliability. The next part of this has not panned out as well however in that this modelling for life time estimation needs to have a defined chloride level at say the first level of reinforcement in order to allow an estimate of corrosion initiation to be made. Unfortunately, as the chloride level measured for initiation of corrosion varies between 0.1% and 6% on real structures, it is not possible to anticipate the corrosion initiation period. Presently the only practical technique for corrosion rate evaluation following initiation is measuring corrosion induced cracking on the surface and this has significant limitations.

There has been some consideration on the affects of reinforcement corrosion on the structural capacity of some of the components of the structure such as beams and columns. Scale models and full size items have been produced and load tested. In most cases with significant corrosion of the reinforcement the structural integrity has been impaired, beams have been particularly affected. It should be noted that in most of the prematurely corroded structures where a decision to replace the component or apply cathodic protection has been made, load bearing degradation has not been a dominant factor.

A large amount of effort has been put into non-destructive testing of reinforced concrete structure with a lot less success than might have been anticipated a few decades ago. As of today, we have surface potential measurement of underlying steel rebar, corrosion cracking and iron oxide staining as our primary diagnosis techniques. Other techniques such as linear polarisation resistance (LPR) have had commercial equipment developed by various companies but, at best, are difficult to interpret in a real-life situation. Acoustic emission has been tested though it has many drawbacks and does not look to be a realistic survey method. Many more techniques have come and gone after a test program has been completed. My favourite was a Japanese technique introduced by Kobayashi (2011) which involved inductive heating of the rebar. In today's energy crisis surveying a building could also warm it for the inhabitants.

We are now much clearer what to do with a structure which is suffering from reinforcement corrosion. Sometimes doing nothing is a reasonable choice, others patching the cracked areas could be sufficient. Another technique might be to replace the structure with a more durable design. A further option is cathodic protection and what we know about this technique is it works well if done properly and then maintained. The converse is also true and in this case it is a waste of money.

REFERENCES

Angst U, Elsener B, Larsen C, Vennesland O, Critical chloride content in reinforced concrete – a review, *Cement and Concrete Research*, 2009, 39, 1122–1138

Kobayashi K et al, A fundamental study on quantitative assessment of steel corrosion by a new NDT method with induction heating, Proceedings of concrete solutions, 4th international conference on concrete repair, Sept 11, 2011, Dresden, Germany

Marcus P, Pour un monde durable: Journee mondiale de la corrosion, Material Tech, April 2020, 108, N1

Index

acoustic emission 56
activation rate control 44
advection 14
alkali aggregate reaction 21, 37

butler-volmer equation 61

capillary flow 14
capillary pores 11
cathedral 4, 6
cathodic protection 32
cementite 15
chloride level 22
cold rolling 16
colosseum 4
corrosion threshold 24, 25
cracks 38
cramps 4
current density 37

debye-huckel 9
debye layer 36
degree of solvation 31
design life 7
diffusion rate control 43
double layer capacitor 48, 49

effusion 31
electric double layer 35, 36
electric field 29
electrochemistry 2
electrochemical processes 1, 2, 3
electrode potential 43
evans diagram 53, 54

faraday 9, 49, 56
faradiac resistance 48
ferrite 15
flash rusting 17
fuel cell 3

gel water 35
grahams law 35
green rust 46

haematite 42, 44
high temperature corrosion 3
hydrogen evolution reaction 51

immunity 47
infinite dilution 31
inhomegeneties 27
interface transitional zone 12

jumps 32

laser induced breakdown spectroscopy 28, 29, 36, 38
linear polarisation resistance 56, 60, 65, 71
load bearing degradation 71

magnetite 42, 44
martensite 15
millscale 17, 42, 43
mixed potential 9
modelling 64

nernst equation 48, 49, 52, 53
nernst-plank model 64, 65

ordinary portland cement 7, 11
oxide films 41, 42

partial saturation 13
passivation 41
pearlite 15
pitting 63
point defect model 45
polarisation curves 53
polyelectrode 53
post tensioned 8
potential field 35
potential gradient 28
potential mapping 65
pourbaix 9, 51
pozzolan 11
pre tensioned 9

reversible potentials 48

sabkha 14
saturated flow 13
silica fume 25
simulated pore solution 44
solid to electrolyte interface 18
solvation sheath 31
steel 7
steel to concrete interface 17, 18
stokes radius 31
strategic highways research program 8

tafel equation 9
tafel plot 52, 54
tafel slopes 60, 61
tortuosity 69

unsaturated flow 13

wüstite 42